Computer addiction?

'Each man to what sport and revels his addiction leads him'

Othello, II ii 6

Computer addiction?

A study of computer dependency

Margaret A. Shotton
University of Nottingham

Taylor & Francis
London · New York · Philadelphia
1989

UK	Taylor & Francis Ltd, 4 John St, London WC1N 2ET
USA	Taylor & Francis Inc., 1900 Frost Road, Suite 101, Bristol, PA 19007

British Library Cataloguing in Publication Data

Shotton, Margaret A.
Computer addiction? : a study of computer dependency.
1. Man. Interactions with computer systems
I. Title
004'.01'9

ISBN 0-85066-795-X Hbk
ISBN 0-85066-796-8 Pbk

Library of Congress Cataloging-in-Publication Data is available

Cover design by Jordan and Jordan, Fareham, Hants.

Typeset in 11/12 Bembo by Bramley Typesetting Ltd, 12 Campbell Court, Bramley, Basingstoke, Hants.
Printed in Great Britain by Taylor & Francis (Printers) Ltd, Basingstoke, Hampshire

Contents

Foreword

Since 1979, a mere ten years ago when the microcomputer first started coming to Britain in quantity, there has been very extensive growth in two particular areas; in the growth of computers and, sadly, of drugs. With the simultaneity of this growth and the intensity with which some people became involved with computers, it is perhaps not surprising that in the early 1980s we began to hear some suggestions of the possibility of 'computer addiction'. The word addiction has been applied to the compulsion of drug-taking since the early 1900s, and indeed it could be argued since 1779 whence the example 'his addiction to tobacco is mentioned by one of his biographers' (quotation from the *Oxford English Dictionary* about Johnson).

In a nutshell one might say that the most interesting and important outcome of Margaret Shotton's doctoral research is to show clearly that the word addiction should not be used about the relatively small proportion of computer users who become intensive computer devotees; or if used, then the term should be interpreted not in the drug sense but more precisely in another version of the definition in the *OED* as 'the state of being given to a habit or pursuit'.

The computer dependent person, to use Shotton's term, is clearly a hobbyist, 'a person devoted to a hobby (sometimes used with a connotation of crankiness)', where hobby denotes 'a favourite occupation pursued merely for amusement or an individual pursuit to which a person is devoted (in the speaker's opinion) out of proportion to its real importance'. This excellent piece of research has shown how the extensive use of a computer can be a most important hobby for some people, not an addiction (in the usual sense of the word) but at one extreme of the very wide range of intense concentration and involvement covered by people's hobby interests.

If the computer, one of the most powerful tools which mankind has so far invented, is never used to cause greater potential distress or danger than as an extreme hobby, we shall have no reason to fear the computer devotee or the computer expert. But that begs a whole different range of research issues!

Professor Brian Shackel
Loughborough University of Technology

Preface

This research was initiated through my combined interests in new technology and in people. As a lifelong observer of the human condition I have always been fascinated in the activities of others, and in trying to determine what makes them 'tick' and brings them fulfilment in life. What is obvious to all is that what holds the attention of one may provide boredom for another. Many in the population feel that chasing a ball around a court or field is a worthwhile and meaningful activity, while to others programming in machine code is infinitely more exciting. Who is to say which is more acceptable?

During a period of four years I was immersed in the lives of people for whom interaction with computers was considered infinitely preferable to the majority of their interactions with people. This is not a belief I personally share, in spite of the fact that I spend much of my life staring at a VDU screen, but one which I came to understand and appreciate fully. People differ in their needs, aptitudes and in their cognitive styles, and happy are they who are able to find an activity which perfectly matches their personality.

The 'computer dependents', who shared their beliefs, their pains and their happiness with me, have enriched my understanding of psychology as no textbook ever could. Their honesty and their ability to lay bear their weaknesses as well as their strengths have proven how dangerous it is to condemn what one does not understand or to show prejudice against those who differ from ourselves.

Early readings about 'computer junkies' and 'hackers' suggested that if I pursued this research I might spend my time with people who were barely human and who were unable to converse with others on any meaningful level. How untrue this proved to be. I met some of the most fascinating people of my life. They were intelligent, lively, amusing, original, inventive, and very hospitable. True, they rarely spend much time communicating with people for reasons explained within this book, but when interest was shown in them and their activities it would be difficult to find more interesting conversationalists. True, many of them were unconventional and unconstrained by society's *mores*, but who would not like the freedom and courage to act without recourse to others? True, some of their relationships were problematic and their activities bewildering and distressing to their partners, but they were no more likely to have failed marriages than the rest of the population.

They were pursuing an interest which not only provided intellectual challenge, fun and excitement in infinite variety, but one which enabled many of them to improve their career prospects considerably. Many used computers not only at home but also at work, and true fulfilment must come to those who are able to combine their hobby with a means of earning a living.

I trust that this book will enable readers to re-evaluate their attitudes to those in society who do not share their own interests, to become more empathetic with those who seem socially inhibited and shy, and to realize that judgment based solely upon observation alone is inadequate when one wishes to understand the machinations of the minds of others.

Acknowledgments

This research formed the basis of my doctoral thesis, and I would like to thank Loughborough University of Technology which funded much of this research. My grateful thanks are also extended to my old colleagues within the Department of Human Sciences and the HUSAT Research Centre of Loughborough University, in particular Professor Brian Shackel and Brian Pearce; to Tom Stewart for all his encouragement and advice; and to the many other friends who showed sustained interest in this work and offered much moral support.

My most fervent thanks are extended, of course, to the computer dependent respondents and their spouses who shared their experiences with me so openly and without whom this work would have been impossible. I trust that although individual differences between the dependents are somewhat hidden by the results from the whole group, that they can each see themselves in the conclusions drawn and that they feel I have explained their situation honestly and factually.

In the light of this research, my final thanks are extended to the old *Commodore 8096-SK* upon which I word-processed the drafts of this book and which, in spite of its age, provided *almost* constant, faithful service!

Margaret A. Shotton
Nottingham
July 1989

Chapter 1
Introduction

Original motivations

My interest in this particular field of the human-computer interaction was stimulated when observing the reactions of a group of students at Loughborough University to their initial introduction to computers. Although most felt enthusiastic initially, this attitude often altered dramatically after their first sessions at the keyboard, and the students seemed to separate into distinct groups. Working with these inanimate machines seemed to provoke various emotional responses, and few remained impartial. Most treated the computer as a useful but complicated 'black box', a tool, which because of their inexperience often seemed to create more problems than it solved; some went so far as to express open hostility, bewilderment and despair; while in contrast a few appeared totally fascinated by the machines to the exclusion of other activities. One of the students appeared to sacrifice most of his work and leisure time in order to program, and became known as a 'computer junkie' to his fellow students.

A little later I heard this term being used by ergonomists who were uneasy about the poor quality of some computer software, much of which appeared to be almost unusable. One opinion held was that such software was designed by 'computer junkies', those who could only communicate with machines and had little idea of users' learning requirements or abilities.

Anecdotal reports, both verbal and in the press, also implied that a wider society recognized such a syndrome. Such colloquial terms as 'computer addicts', 'machine code junkies', 'microholics', 'hackers', and 'micromaniacs' were used to denote people who were considered by others to use computers excessively. Such reports suggested that this apparent dependence upon, or 'compulsive' use of, computers could have deleterious consequences, especially upon the social relationships and academic achievements of the young. However, in spite of such beliefs little or no serious research appeared to have been carried out to investigate this syndrome.

Press reports which had alluded to the existence of computer dependency were sought, but few could subsequently be traced. This probably occurred because such items seemed to have been confined to the tabloid and the popular computer press which were not well represented within the abstract indexes examined. However, those which existed indicated that there appeared to be a common acceptance of the syndrome of computer dependency. The reports, most of which were apocryphal in nature, suggested that some people were becoming 'hooked' or 'addicted' to the microcomputer in much the same way that others become dependent upon certain drugs.

Examples of the more serious of the press articles included those quoting Reynolds who in 1983, through various newspapers and journals, expressed some concern about a few undergraduates whom he felt were becoming excessive users of computers. He suggested that such students tended to be introverted isolates, who found safety in their interaction with the computer but made very poor students, often failing their examinations because of their poor work methods. Pearce, from Loughborough University, also believed he had encountered this phenomenon whilst working in the computer industry:

> 'The relationship with the machine can take over to the point where individuals are better able and prefer to talk to computers than to people. They will log on almost before talking to anyone.'
>
> Pearce (1983)

As a first foray to investigate this area I visited the local branch of a large newsagents, a video game arcade and a computer users' club to make observations about the growth and influence of computers, their games and literature. The manager of the newsagents seemed convinced of the existence of computer dependency, and stated that he had found it necessary to reorganize the layout of his store in order to cope, not only with the multitude of new computer magazines, journals, hardware and software being sold, but also with the different type of customer being attracted to these goods. As computer magazines began to outnumber women's magazines, previously the most popular, he had assigned them to their own area. This was found necessary as the computer enthusiasts hindered other shoppers, as they tended to use the shop as a library where they could browse for hours and where they could meet and exchange information with others of like mind. He felt that Saturday shopping expeditions had also changed in nature. Where all family members had previously shopped together, now some were being left at the newsagents to be collected an hour or so later. The manager stated that this arrangement usually entailed the wife continuing with the weekly shopping while husbands and sons were left to play with the hardware and read the computer magazines. He believed that many customers appeared to devote all of their spare time to computers and cited frequent occasions where children, usually boys, spent hours in his shop when they should have been at school. Similar situations were thought to occur in other shops selling computer equipment.

The amusement arcades appeared to have quickly adapted to the new enthusiasm for the fast-moving, computerized video games and such games appeared to attract players who would often spend many hours perfecting their skills. The players, once again mainly male, would vie for superiority of scores, and if novices would frequently spend a great deal of money trying to perfect their skills. The games experts tended to be viewed unfavourably by the arcade owners, as when skilled they were able to play one game for long periods of time, spending little money but monopolizing the machines. Some of the players admitted they were 'hooked', and although most wished to own their own computers so that they could play without spending money they enjoyed the prestige earned while playing with others.

The computer club I visited seemed to be very disorganized and appeared mainly to act as an exchange for computer magazines and pirated software. The club members seemed convinced that people did become dependent upon computers, and two admitted that using computers had come to dominate their waking hours to the exclusion of other activities.

All of the conversations undertaken, whether with colleagues or computer users, seemed to suggest that the press were not acting irresponsibly when reporting the existence of the syndrome of computer dependency. Those who recognized it in others tended to see it as harmful, while those who confessed to being 'hooked' on computers saw it as a most exciting and pleasurable activity which could only be of benefit to their education in new technology.

I decided that research into this area would not only prove fruitful but also worthwhile. Many leisure activities have merely been fashionable fads lasting only a few months (the Rubik cube, the skateboard and the hula hoop are examples), and no doubt there have always been a few concerned members of society who suggest that such intensive interest could prove harmful. Past experience has shown that when these activities are mastered interest in the activity often dies very quickly, and with it the concern. However this may not occur with computing. The computer is a versatile tool which can provide an almost infinite range of activities to the user, and also gives the impression of being an intelligent entity which some appear to endow with anthropomorphic characteristics. For these reasons it was likely that personal computing could become a long-term activity for some, although it was unknown whether its influence could prove any more damaging to them than merely a drain on their finances.

The research was established and centred upon the people directly involved with this modern syndrome; the individuals for whom the computer had come to dominate their lives and interests. The investigation aimed to determine if and why an interest in computers develops to the extremes suggested by some people; to discover who was and who might be affected; to investigate whether certain factors could be isolated as instrumental in the development of such a syndrome; and to attempt to establish which needs were being fulfilled by using computers in a fashion to cause concern to others.

Without wishing to disappoint any computer enthusiasts who may read this book, the main thrust of this research does not concentrate upon the computers, the programs, the bits and bytes, the ROMS and RAMS, or any other artefact of new technology. Nor is it specifically concerned with those who become 'addicted' to video games machines in arcades; this is a separate issue which involves some element of gambling, and where the computer is merely a vehicle for the game and rarely of intrinsic interest in itself. In addition, the effects of new technology and artificial intelligence upon society as a whole will not be discussed at length; this I leave to philosophers, who are far more learned in this area than I. Neither is this research solely concerned with the deleterious effects of computer dependency upon the individual and others, as this appears to overlook the more fundamental issue of why computer dependency should occur in the first place.

Although I was convinced that there was a need to study this area of the human-computer interaction, further exploratory investigations were necessary before the full research project could be designed. Too little was known about this modern syndrome to begin formulating hypotheses or making conclusive decisions about methods to be used. Also, before embarking on the research it was felt necessary to select terminology which could describe this syndrome in less emotive terms than are used by the general public.

Definitions used: computer addiction *or* dependency

I experienced great difficulty when looking for a term which could be used to describe adequately and accurately the nature of this type of human-computer interaction. The anecdotal reports had freely used the terms 'addiction', 'compulsion', and 'obsession' to define the condition of extensive computer usage, but none were considered to be strictly appropriate.

A *compulsion*, as defined by the *Supplement to the Oxford English Dictionary*, (Burchfield, 1976) is 'an insistent impulse to behave in a certain way, contrary to one's conscious intentions or standards'. This notion is substantiated by the definition in the *Dictionary of Psychology* (Drever, 1952) which describes it as an 'irresistible inner force compelling the performance of an act without, or even against, the will of the individual performing it'. However, there had never been suggestion in any of the reports that this computing activity was being undertaken against the will of the participants; indeed it appeared to be quite the contrary. The computer users who took part in this research invariably stated that their computing activities were undertaken voluntarily and willingly, rendering this definition unsuitable in this instance.

The *Supplement to the Oxford English Dictionary* (Burchfield, 1972) defines an *addict* as 'one who is addicted to the habitual and excessive use of a drug'. Despite the obvious circularity of this argument, all other definitions and uses of the term 'addiction' in the current psychological literature only use this term with reference to substances ingested by the body. The term 'addiction' was therefore also unsuitable.

Similarly the term *obsession* proved inaccurate as its use refers only to thoughts and not to actions. The *Oxford Dictionary* (Burchfield, 1982) defines an obsession as 'an idea or image that repeatedly intrudes upon the mind of a person against his will and is usually distressing'. The *Dictionary of Psychology* sees it as 'a persistent or recurrent idea, usually strongly tinged with emotion, and frequently involving an urge towards some form of action, the whole mental situation being pathological'. This was inappropriate for the majority of computer users. There was no doubt that the thoughts of some were dominated by computing when away from the machine, but the physical process of computing was of equal, if not greater, importance to them than the ideas engendered.

Orford, in his book *Excessive Appetites: A Psychological View of Addictions* (1985), became engrossed in a discussion of terminology. He also found 'addiction' unsuitable for his purpose, in spite of the fact that he and his publishers finally resorted to the use of this word within the book's title. Orford preferred the words 'excessive' and 'dependency' which he employed throughout his text. However, the word 'excessive' is very subjective: excessive in whose eyes! The research participants did not find their time at the computer excessive, but rather wished there were more hours in the day in which they could undertake their chosen activities.

The more objective word *dependency*, used frequently by Orford, was found to be the term favoured by the *Penguin Dictionary of Psychology* (Drever, 1952), even when describing drug use.

> 'In recent years the term (dependency) has come to be favoured over the terms addiction and habituation in scientific writing. Put in its simplest terms, an individual is said to have developed dependence on a drug when there is a strong, compelling desire to continue taking it.'

Similarly the World Health Organization (1964), because of the confusion in accurately defining various conditions, had abandoned terms other than 'dependency' for all situations. It was believed that 'dependency' neither implied a particular level of tolerance, habituation, side-effects or withdrawal symptoms, but was open to many interpretations and depths of severity.

From these arguments it was decided that the term 'dependency' would be the definition most suitable to be employed within this study. However, 'dependency' is not a definition commonly used by lay people, and a popular term was needed which would convey an equivalent meaning when making contact with the general public. As has been mentioned, 'code junkies', 'computer addicts', 'microholics', 'hackers' and 'micromaniacs' are all colloquial terms which have been used to describe persons who are considered to be dependent upon computers, but these were considered to be unsuitable because of the negative, emotive connotations associated with them.

Discussions with computer professionals, psychologists and computer users yielded a term which did appear to fulfil the needs of this research. The expression *to be hooked on computers* was felt to imply the essence and condition of dependency, but managed to avoid the very negative, emotive allusions denoted by the other terms. This also appeared to be the term most frequently and popularly employed by computer users in common speech to represent computer dependency, and was therefore the term used when soliciting people to take part in this research.

The term 'sydrome' is also used within this book to encompass the phenomenon of computer dependency. The concurrence of symptoms constitute a syndrome and it was the configuration of these symptoms, among other aspects, which was of interest during this research.

Chapter 2
Initial investigations: Does computer dependency exist?

Do professionals believe in the existence of computer dependency?

Before any major research was undertaken, investigations were carried out to determine whether the syndrome of computer dependency had been identified by other professionals and, if so, whether they had considered it to be an area of concern. It was intended that by systematic enquiry of those who could be considered to have some knowledge and experience of the area that the more apocryphal stories describing the characteristics of this syndrome be accepted or rejected. Further investigations were also undertaken to contact individuals who personally considered themselves to be dependent upon computers in order to gain an initial impression of the situation. The intention at this stage was to determine the extent to which people believed that computer dependency was identifiable, rather than to conduct deep, statistical analyses. Thus contact was made with individuals and organizations whom it was thought would have either direct or indirect knowledge of the syndrome of computer dependency if it did exist.

Information was obtained from four sources:

- Existing literature;
- The authors of the literature and other psychologists;
- Computer studies teachers;
- Professional care agencies.

The literature review aimed to uncover references about computer dependency, to ascertain the current opinions of this syndrome and to discover if any previous research had been undertaken to investigate this aspect of the human-computer interaction. Correspondence was initiated with the authors of the literature to try to establish the basis and reasons for their comments, and with other psychologists either because of their general concern for the human condition or for their specific work with new technology. Contact was also made with teachers of computer studies who might have taught computer dependent individuals, and with professional care agencies who might have offered advice to people if computer dependency became severe enough to warrant their intervention.

Existing literature

The earliest reference to computer dependency I found was made in 1970 by two far-sighted individuals.

> 'The interest and excitement that is likely to be stirred up by being able to dial up and work with an ever-growing number of computers is likely to be far greater, and far more time-absorbing than the attraction of hobbies such as ham radios . . . perhaps the majority will fall under the narcotic spell of programming. Working on ingenious programs is, to a certain type of mind, endlessly captivating . . . whether this will come about in the next ten or fifteen years is difficult to tell, although we tend to doubt it.'
>
> Martin and Norman (1970)

Conforming to Toffler's vision (1971) of a society unable to comprehend the speed with which technological innovations would develop, Martin and Norman, although unable to envisage the advent of the personal computer or to estimate the year in which it would be marketed, were sensitive to the possibilities that people could become totally fascinated with computer programming.

Very little since that time was found to have been written about computer dependency. To 1984 that which existed had been contributed by psychologists or computer scientists usually describing main frame computer users within universities. In general the authors were forceful in their opinions and often very scathing towards those they described as dependent upon computers or video games. However, even these reports appeared to be little more than apocryphal tales, similar in content to the comments seen in the press, with none being substantiated by empirical research.

Joseph Weizenbaum, Professor of Computer Science at the Massachusetts Institute of Technology, implied that he recognized the syndrome and was the only author to have dwelt at length upon this subject. In his book, *Computer Power and Human Reason* (1976) he was pessimistic and damning about the people he described as 'compulsive programmers' and devoted one complete chapter to this subject. He described such people thus:

> 'Bright young men of disheveled appearance, often with sunken glowing eyes, can be seen sitting at computer consoles, their arms tensed and waiting to fire their fingers, already poised to strike, at the buttons and keys on which their attention seems to be as riveted as a gambler's on the rolling

> dice. When not so transfixed, they often sit at tables strewn with computer printouts over which they pore like possessed students of a cabalistic text. They work until they nearly drop, twenty, thirty hours at a time. Their food, if they arrange it, is brought to them: coffee, Cokes, sandwiches. If possible, they sleep on cots near the computer. But only for a few hours—then back to the console or the printouts. Their rumpled clothes, their unwashed and unshaven faces, and their uncombed hair all testify that they are oblivious to their bodies and to the world in which they move. They exist, at least when so engaged, only through and for the computers. These are computer bums, compulsive programmers. They are an international phenomenon.'

Weizenbaum had few reservations and believed computer dependency to be a compulsion; an all-consuming drive which could totally dominate lives. He saw the computer acting as a challenge to the programmers' power rather than to their intellect, and believed that their aim was to make the computer bend to their will rather than to use it as a tool. He stated that the compulsive programmer was 'merely the proverbial mad scientist who has been given a theatre, the computer, in which he can, and does, play out his fantasies'.

Weizenbaum differentiated between 'compulsive' and 'professional' programmers by suggesting that the former spend their time producing aimless, lengthy, overly-ambitious, ill-documented programs which only they could understand and maintain, while the latter were solely concerned with problem-solving through the use of the computer, using the keyboard only after considerable preparatory planning had been undertaken. While the professional was an efficient member of a computer centre's team, the compulsive programmer on the other hand was described as a person whose main aim was to have the opportunity merely to interact with a computer and rarely produced useful output.

By 1984 Weizenbaum expressed further anxieties about dependency upon new technology, this time with reference to arcade and computer games. He believed that violent computerized games could engender 'psychic numbing'; the emotional distancing of the players from the results of their actions. Unlike doses of violence from a television screen, he felt that computer games were far more destructive because of the participatory nature of the interaction.

Computerized games have long been criticized as leading to violent behaviour and 'addiction', to the extent that the Philippines and Singapore have both banned video arcade games, claiming that they cause aggression, truancy, psychological addiction and stealing in order to support the habit. Similar thoughts have been mooted in other countries, including Britain. In 1981 a Bill was laid before Parliament at Westminster entitled the *Control of Space Invaders and other Electronic Games* although it was subsequently defeated (Parliamentary Debates, 1981).

Conflicting opinions surround the reasons why some children appear to become dependent upon such games. Loftus and Loftus (1983) defended video games for their educational value, but believed that dependency upon them was due to the inherent characteristics of the games. Favaro (1982), although unconcerned with the violent aspects of video games which he considered had always formed an important part of children's play, believed that the game was blameless for any subsequent 'addiction'. He felt that such children already had existing emotional problems which would manifest themselves in some type of compulsive activity or another.

McCorduck (1979) saw computer dependency also as an artefact of personality, but assumed that it was from 'such obsessive and badly balanced persons that much of humanity's treasure has come', qualifying this by stating that nobody would have thought that living with Beethoven 'was a picnic'. It is not difficult to draw similarities between the dedication of the computer enthusiast and that of the musician, artist or writer, and machines which mimic or extend human intelligence are perhaps more likely to create involvement and attachment than other artefacts, as suggested by Shallis (1984).

> 'Machines that "talk back" hold a fascination, almost hypnotic, that many people just cannot resist. The anthropomorphized machine really can evoke an emotional response from people that is hard to distinguish from genuine human feelings.'

Recognizing the possibilities of emotional attachment by a user to a computer, Boden believed it to occur because of the feelings of power and control that the computer could engender.

> 'The computer is a toy of quite a different order from any previous plaything, in its ability to satisfy—or to appear to satisfy—various deep-seated human needs. It offers us the illusion of total control, and it appears to give us its full attention, often being immediately responsive to every remark addressed to it . . . these flattering responses are rarely found in human beings, and not reliably even in dogs. If we like to appear a Napoleon to our four-footed friends, how much more we may relish appearing a Zeus to our computer.'
>
> Boden (1981)

A more ominous thought came from Thimbleby (1979) who suggested that the compulsive programmer's need for power over the computer would extend to a desire for power over other people. He interpreted this behaviour 'as a neurotic struggle to gain power over the father figure', or conversely as a regression 'back to the womb'. However, Frude (1983) recognized not only the need for control and power but also the more light-hearted aspects of play and exploration, which he considered to be legitimate by-products of some computing, as those which could lead to excessive use.

Observations of the behaviour of students with computers led Zimbardo to believe that 'fascination with the computer becomes an addiction, and as with most addictions, the "substance" that gets abused is human relationships' (Zimbardo, 1980). He obtained and published printouts from computer network communications between students who discussed their computer dependency and the advantages and disadvantages of their activities (*The Hacker Papers*, 1980). Of the seven students quoted, two gave the impression of wishing to escape the excesses of their dependency. They described a syndrome which created antisocial behaviour, inhibited students from taking part in university activities and deleteriously affected their studies. These charges were confirmed by the remaining students, but were defended by the fact that they considered computing to be an excellent substitute for poor human relationships, while acting as a stimulating, intellectual challenge.

Although some of the examples in the literature made the situation of dependency upon computers sound dire, it is necessary to bear in mind the possibility that some might have been written in a deliberately provocative manner in order to draw attention to the subject matter.

Personal views of authors and other psychologists

In response to my letter, Zimbardo revealed that he was deeply concerned about the students who appeared to be dependent upon computers, and felt a serious investigation into the 'social dangers of excessive computer use (addiction)' was warranted. His own research centred upon the phenomenon of shyness and he felt that some students used computers as an escape from social interaction. He admitted that his papers were aimed to be deliberately 'alarmist', in order to try to 'raise the consciousness' of the general public about the negative effects which he felt could develop from this type of interaction with computers. This deliberate desire to be provocative perhaps illustrates the concern felt by some who have observed this syndrome.

Further communications revealed that Margaret Boden, Professor of Philosophy and Psychology at Sussex, now concentrating her research upon artificial intelligence, believed the condition could exist and felt this topic to be of great social importance; she was particularly concerned with the apparent sex differences observed in computing generally. Hans Eysenck, Professor of Psychiatry at London, considered such people to be suffering from neuroses, simply finding solace in the computer. Donald Broadbent, at Oxford, believed that computers were anthropomorphized by some people, but both he and John Nicholson, London, saw this human-computer interaction as more positive than most. They felt that the computer was used as an escape by some people, often from poor social relationships, but they believed that computing could relieve some of the frustrations of life while adding meaning and control.

These initial investigations indicated to me that a few computer professionals and psychologists were concerned that the use of computers could alter the behaviour patterns of some people, and the literature review suggested that a small proportion of the computing population could be described as *computer dependent*.

Informal interviews with computer studies teachers

Interviews were conducted with teachers from computer centres at Loughborough and Brunel universities and within two Leicestershire secondary schools. All the interviewees believed that the condition of computer dependency existed. However, they felt that the situation was not widespread and was therefore not considered to be a serious problem for the majority of computer users.

The Loughborough University staff reported that facilities existed which enabled them to curtail the amount of time students spent on the mainframe computer; facilities which were utilized for financial reasons rather than because of any effects upon the individuals. This did not necessarily prevent students from spending most of their time computing though, as the staff believed that such users also owned their own microcomputers. (Staff it seemed were uncontrollable and left to their own devices.)

Reynolds from Brunel University, who had frequently been quoted in the press, confirmed many of the points raised in the literature. He felt that the students who showed symptoms of computer dependency also experienced social problems, some of which could be relieved by gaining social recognition for their skills. He stated that some of their programs were 'unreadable, lengthy and inefficient' and often would not run when tested, and that a few of the students performed very poorly in computing examinations in spite of the time they spent at the keyboard. Such students, when recognized, were limited to only five hours per week on the mainframe computers or risked losing coursework marks. Because of the students' poor performances, this department had devised a questionnaire to be given to all candidates applying for computer studies courses, with the aim of eliminating potential dependents through analysis of their interests and previous computing experience. Reynolds felt that industry needed people who were socially competent and could share ideas, not those who were only interested in exploring computer systems.

Within the two schools visited, computer staff also recognized the syndrome and thought that this could sometimes be detrimental to the children's other studies and social lives. It was salient to note, however, that the teachers interviewed appeared positively to encourage the computing interests of these children, virtually giving such pupils free access to the computer rooms. One gained the impression that these few special children had become the protégés of the computer staff. One teacher personally acknowledged the situation, stating that 'the powers of addiction are recognized by anyone who uses a computer', although he did admit that some of the children, particularly the girls, seemed to be totally disinterested in new technology.

Contact with professional care agencies

The communications with the professional care agencies bore little fruit. Most felt unable to divulge information, and of those who would express an opinion none had direct knowledge of computer dependency or of help being given to anyone with this syndrome.

Although the characteristics associated with computer dependency appeared to be manifest in some people, at that time it seemed that neither they nor their families were turning to the care agencies for advice.

Personal contact was, however, made with the Leicestershire chief educational psychologist who indicated that such a syndrome would probably need to be quite severe and widespread before his team would become involved. He admitted that it was a number of years before the effects of solvent abuse had been brought to the notice of educational psychologists, and he felt that the presence of national publicity about 'glue-sniffing' had made teachers more aware and more likely to refer children to his department.

Preliminary analysis

One may conclude from the literature review that there appeared to be a number of people, recognized by academics and computer professionals, showing behaviour patterns which might be described as those of computer dependency. Much of the literature was American and most concerned students or university-based computer scientists working on mainframe computer terminals. However, it is questionable whether these examples can be considered to be representative of the population of computer dependent people. Observers of the human condition, with the time and resources available to them in order to publish such observations, tend to centre within the seats of higher learning, namely the universities. Undergraduates have made suitably captured subjects for many a study, and this literature may merely reflect this trend.

The authors provided diverse and often contradictory theories for the causes of computer dependency, from the need for control and power to a desire for intellectual stimulation. The computer was described as being more than a mere tool, being bestowed with anthropomorphic characteristics which could provide positive reinforcement and satisfaction. Few viewed the effects of such extensive use of computers positively with the behaviour described as compulsive and addictive, although one author suggested that original creativity could result from this type of human-computer interaction.

Although none of these writings was based upon empirical research specifically designed to look at this area of human behaviour, and provided little of a substantial nature upon which further research could be solidly based, the results helped to confirm the belief that the area was worthy of further study. (Since this literature review, further references to computer dependency have been published, details of which are discussed within later chapters).

Similarly, the interviews carried out with the computer studies teachers showed that they also recognized the condition of computer dependency amongst some of their users, citing signs which corresponded to those described in the literature. Most considered that the phenomenon occurred to those with social difficulties, while others believed it could create social and academic problems especially where students were concerned.

Although the professional care agencies reported no knowledge of help being given to computer dependent people, this did not necessarily mean the problem did not exist. However, it could be assumed that the syndrome was less serious than had been suggested in the literature.

The opinions which were offered during these exploratory forays into the area could only be taken at face value and could not be seen as objectively true or as hard evidence for the existence of the syndrome. However, there appeared to be sufficient grounds to continue the investigation in a more formal manner. At this point no contact had been made with individuals who considered themselves to be dependent upon computers and the next phase of the research involved tracing such people to discover the characteristics exhibited.

Do individuals believe they are personally dependent on computers?

The research to this point had revealed that various professionals believed in the existence of computer dependency and that a few American students described themselves as such (*The Hacker Papers*, 1980). In order to discover whether any British computer users considered themselves to be dependent, efforts were made to contact and talk to them directly.

Undergraduate students

Although not wishing to extrapolate from a university population to a wider one, it seemed sensible at this stage of the research to enlist the cooperation of students. They were available, and earlier discussions with the staff of the computer centre at Loughborough University indicated that there were a few students on the campus who could be considered to be dependent upon computers. Five such students were contacted and interviewed.

The aim of the interviews was to investigate the students' computing habits and attitudes to see if these bore any relationship to the opinions offered by the literature, and to discover whether the students considered themselves to be dependent upon computers. The interviews also covered such topics as schooling and academic achievement, hobbies and interests, together with details of their family and social lives. The initial investigations implied that computer dependency arose either as an escape from social contact or was a contributory factor towards it, and the interviews aimed to throw some light on these theories by examining the personalities and needs of the students through self-report. No detailed research was undertaken at this stage, and because of the small and biased nature of this sample no standardized personality tests were used.

These first interviews yielded considerable qualitative data. Very simple content analysis carried out on the results indicated that there were numerous similarities between these five students, not only in their computing habits but also in other aspects of their lives. This was perhaps hardly surprising as they were all male, of much the same age, and were studying for degrees at a university of technology. However, their responses did indicate that they exhibited many of the characteristics previously ascribed to computer dependency in the literature.

The students recognized and believed that computer dependency could occur, and all five considered themselves to be 'hooked' on computers describing themselves as 'computer junkies'. They spent an average of about 30 hours of their spare time each week on the computer. Two students had spent over 15 hours at the keyboard without a proper break for food or rest, and ten-hour sessions were not uncommon. All had attempted to ration themselves to a certain extent, and they stated that they often lost all track of the passage of time when at the keyboard.

The interviews gave a great deal of insight into the students' activities at the computer, much of which, but by no means all, had been anticipated. All of the students owned at least one microcomputer but despite their great interest in computing none considered themselves to be very efficient programmers. They very rarely planned their programs, preferring to create 'hands on', allowing the program to develop as they worked. They stated that this method often meant that they had to spend many hours 'debugging' their programs of faults, and it was of some surprise to discover that debugging was the activity from which they tended to derive their greatest pleasure. They enjoyed chasing faults and mistakes, and spent much of their time elaborating the programs, speeding them up or improving the graphics. These views confirmed the observations made by Weizenbaum (1976), who suggested that computer dependent programmers preferred to interact with the computer rather than produce functional code.

Computer games were sometimes played, and the students particularly enjoyed the elaborate adventure games which they considered to be more intellectually stimulating than the arcade-type games. Also popular was 'hacking', illegally gaining access to computer systems via the networks, which was considered to be a highly prestigious activity open only to the skill of the 'real junkie'.

The students tended to look upon the computer not merely as a tool, but also as a toy or a learning machine. Its use appeared to have become an end in itself, and the computer was rarely used for solving specific problems unless these concerned learning about the computer or its languages. The aim was 'to find out what it could do and make it do it'. Power and mastery of the system did seem to be important, as had been suggested by Boden (1981). This is a facet of the human-machine interaction which is perhaps unique to computers, as it is difficult to imagine people wishing to dominate their cars or stereo systems, however much they 'play' with them. Three students also tended to anthropomorphize their computers, viewing them as 'partners' and 'friendly colleagues' who were 'easy to get on with'.

Two of the students were performing badly academically, achieving grades which did not reflect their earlier successes at school. They believed that their time spent computing was detrimental to the other demands of their courses, but did not consider this to be a serious problem. They were enjoying their computing and felt confident that they would get a degree, even if it was not first class. The other students seemed able to find some time for their other studies, although this was often achieved by adapting as much of their work as possible so that it was computer-based, or by choosing options which incorporated computing.

Although preferring to work alone they did obtain satisfaction from being able to help others with computing. Any feelings of being taken advantage of were outweighed by the increase in prestige and social gain amongst their peers. They thought others saw them as introverted, lonely eccentrics, which they did not dispute, but when asked to make comparisons with their fellow students they described themselves as more mature, more intelligent and logical.

Perhaps confirming Zimbardo's suspicions (1980), the interviews revealed that all five students found social interaction difficult and had done so from an early age. Despite their academic successes at school they did not enjoy those years, as they had tended to feel ostracized and different from other children. They felt intellectually superior to their peers and that they had little in common with them, leading to unhappy isolation in some cases. These feelings had not diminished while at university, in spite of the fact that they were mixing with many students of similar intellectual status. Only one student felt that he might have made more effort to socialize if computing had not been his dominant interest; the others felt that they would have been undertaking other hobbies or doing more coursework as an alternative to computing.

There were also surprising similarities between their preferred hobbies and interests. They were all interested in reading, especially non-fiction and science fiction, in mechanical, electronic and electrical hobbies, and in mental games and puzzles of all sorts. They admitted having very little interest in either the social or sporting activities offered by the university, and did not consider themselves to be in any way artistic, although listening to music was popular. However, since computing had become the dominant interest they rarely found the time to practise many of these old activities.

Details of the students' family backgrounds tended to be described as stable and middle-class. Their parents did not seem to be out of the ordinary, although fathers tended to be described as unloving or reserved. Mothers were either seen in a similar light, or described as very loving, caring and devoted.

It is difficult to draw any firm conclusions based solely upon interviews conducted with a small sample, which, by its very nature, was biased. However, all five students considered themselves to be computer dependent and appeared to treat computers differently from the majority of users (who view them merely as tools). They were not only using the computer for more hours than most students, but they held different attitudes towards them and used them for different activities. The process of programming seemed more interesting to them than the end-result, and power and mastery over the machine proved important. If computer dependency can be measured by the time spent upon the computer and by the attitudes towards it, one may conclude that these male students were indeed all dependent. Their lives were said to be totally dominated by computing, with every spare moment being spent at the keyboard.

Although there was great concordance between the reports from these students and those found in the literature, all the people described were either students or computer-centre workers and unrepresentative of the general population. One cannot therefore assume that they are similar to others in the community who may also be considered dependent upon computers. However, frequently recurring characteristics were apparent within these computer dependent samples which may be described as operational definitions which could be behaviourally and observationally identifiable. These included:

- A particular personality type, usually described as introverted;
- Excessive amounts of time spent using and thinking about computers;
- Computing undertaken for its intrinsic merit;
- Programming often without a definite, useful end-product;
- Programs unstructured, poorly written and ill-documented;
- Enjoyment of debugging and refining programs;
- Need for power and control over the computer;
- Computer interaction used as an escape from other relationships;
- Lack of desire or time to take part in previous activities;
- Detrimental effects upon academic work.

Although a list of operational characteristics could prove useful during the research, they were not sufficient to form the basis of significant hypotheses at this stage, and the aim was always to extend beyond these factors to gain a solid body of knowledge which could be used to describe computer dependency.

The investigations carried out suggested that computer dependency did exist, albeit only to a small section of the community. Further work was initiated to attract people from the rest of the general population with whom the research could be continued.

The general public

The initial surveys had indicated that professional agencies were not, as yet, receiving referrals from people who could be considered to be dependent upon computers, therefore it was impossible to obtain research participants from pools of clients seeking advice. It was also considered undesirable to concentrate further upon the biased population within universities, so volunteers from the whole population of Great Britain were sought, with the aim of gaining as wide a perspective of the syndrome of computer dependency as possible. Although volunteers do not make ideal research subjects because of the biases inherent to them, volunteers had to be used in order to obtain a satisfactory panel of subjects. Any participants gained as volunteers from the general public would form the subject panel for all future investigations, and subsequent results should be viewed as being drawn from the population in this manner.

In order to eliminate as much bias as possible, the methods used to attract volunteers had to be of a type which could be brought to the attention of as many sections of the general population as possible, and the most comprehensive method of achieving this aim appeared to be through publicity in the national press. A press report was prepared which contained a description of the proposed work and which asked for those who were directly or indirectly familiar with computer dependency to respond if they were willing to take part in the research. Anonymity was assured. This was sent to various national newspapers and magazines, placed on one national and three local computer networks, and sent to the more popular computer magazines and journals. The publicity also led to interviews being given to numerous national and local radio stations, further increasing the coverage of the publicity.

It was difficult to anticipate the level of response which would be achieved by these impersonal methods. Computer dependency seemed to affect very few computer users and it was felt unlikely that many of those who considered themselves dependent would wish to volunteer to take part in such research. No financial incentives were offered, the publicity relied merely on the goodwill and interest of people to respond. However, over a period of months 180 responses from the general public were forthcoming by letter, telephone or via the computer networks. Only six were from people describing the dependency of others, the vast majority were self-reports from people who either considered themselves to be dependent upon computers or were merely interested in the research.

All those who responded to the publicity were sent a very brief questionnaire asking them to state whether they believed themselves to be dependent upon computers, using the colloquial term 'hooked on computers'; whether this caused problems to themselves or others; and to estimate the length of time each week that they spent on their computers at home and at work. This introductory questionnaire yielded a total of 151 responses, a response rate of 84 per cent. 121 considered themselves to be dependent upon computers, and all further results are derived from this group. Of them, 95 also stated that they were willing to be interviewed.

The results (Table 2.1) revealed that all but one of the respondents used computers at home and 69 used them at work. The average time spent on the computer at home or at work was over 20 hours per week, with those computing in both situations showing a mean of over 40 hours. Nearly one third of the group stated that their computing caused problems to themselves, while almost half believed it caused problems to other people.

Table 2.1 Summary of results from the introductory questionnaire

Reported time spent computing (hours/week)	N	Mean	S.D.	Problems caused	Yes	No
At home	120	22·4	12·3	To self	38	83
At work	69	21·6	17·1	To others	54	67
At home and work	68	42·8	20·4			

Although the literature seemed to suggest that most computer dependency occurs amongst people who use computers at work, especially within large computer centres, fewer than expected (57 per cent) used computers in the work situation, while all but one used a home computer. No doubt the working environments of the authors accounted for the biases seen in the literature, although Reynolds, during his interview, mentioned that he felt some students came to university already dependent because of their home experiences, and the interviewed teachers thought that some children had become dependent upon their home computers before becoming heavy users at school.

Although Klemmer and Snyder (1972) indicated that people were very inaccurate when estimating the times they spend on various activities, the results from the Introductory Questionnaire revealed that many of the respondents who described themselves as dependent did indeed believe that they spent long periods of time on computers. One man, whose letter revealed him to be unemployed, reported spending 72 hours per week on his home computer, and over a quarter of those computing both at home and at work estimated that they spent more than 60 hours per week computing.

In all instances the estimated time spent at the computer showed extremely high variance within the group of respondents. The higher variance observed in those using computers at work was accounted for by the fact that most of them either used the computer full-time, for 35 or 40 hours, or merely for a few hours per week. The results from the estimated times spent at home could be interpreted to show an average of about three hours per day for each person. It is therefore possible to see that if one's evenings are spent computing then problems could be caused to other people; a situation reported by nearly half of the respondents. However, it is less clear at this stage how computing can cause personal problems when it is a voluntary activity; although if, as with the students' academic work, it encroaches upon other activities this is more understandable.

Preliminary conclusion: Confirmation of the syndrome's existence

The initial investigations aimed to establish if there was *prima facie* a good case for the existence of the syndrome described here as computer dependency. Although the methods were chosen for their economic use of time and money they aimed to be as varied and as comprehensive as possible. In all instances where personal contact was made with people who worked with computers and computer users there was belief in the occurrence of this syndrome. Some expressed concern that the excessive use of computers could lead to negative influences upon academic and professional work and upon social interaction and relationships; theories which appeared to be confirmed by the students who had been interviewed. None, however, considered it to be a widespread problem, rather believing that it affected just a few individuals amongst the vast numbers of computer users. The establishment of the existence of the syndrome of computer dependency appeared to be confirmed.

As with other dependencies, upon alcohol or drugs for example, it is not easy to determine fully the extent to which a person is dependent upon computing. If some use their computers for only 10 hours per week but claim to be dependent while others use theirs for 40 hours per week and claim not to be, one can at this stage only accept their evaluation. None of the respondents who claimed to be computer dependent were barred from taking part in the research, even if their computing times did not seem particularly high. They described themselves as dependent upon computers and this was considered sufficient to warrant their inclusion within this exploratory study.

The attitudes towards the computer and the computing activities undertaken must remain the factors which determine the condition. The respondents' dependency cannot therefore be proven, but this does not preclude its existence. Other authors have felt free to write about this condition; a condition also recognized by many of the general public. What can be confirmed is that all appeared to be talking about the same condition; the need some people have to use computers in a manner which is usually extremely time-consuming, often with little thought to the output of that computing, in a manner which seems to suggest an interaction difference from that with other artefacts.

The remaining chapters of this book detail the results obtained from research undertaken to discover more about the activities, attitudes and personalities of those who considered themselves to be dependent upon computers.

Chapter 3
An overview of the methods used

The survey and psychometric test methods employed

This research used personal interviews, questionnaires, scales and inventories as data gathering techniques, and repetition of responses across the various measures aimed to increase the validity and reliability of the data obtained. Verbatim quotations made by the interviewees have also been incorporated into the results to add validity and to eliminate interviewer bias in the interpretation of the data.

Interviews were selected as they gave the opportunity to obtain extensive data, covering many areas of interest at varying levels of complexity within the framework of one medium. In addition, the use of interviews as exploratory, measuring instruments also facilitated the identification and study of the relationships between variables and the formulation and testing of hypotheses as they arose. By listening to the interpretations of those who had actually experienced the behaviour in question, personal interviews proved the ideal means for gaining an understanding of this new area of study. The data from the interviews were also used for the formulation of *questionnaires and scales* to be completed by all participants (whether interviewed or not) from which more objective, quantitative data could be obtained.

The qualitative interview data were subjected to simple content analysis to determine the relevant variables and their variability and to identify the range of issues, behaviours and attitudes found. From this the questionnaires and scales were constructed, the data from which were used to increase confidence in the results, to give more powerful and convincing explanations, and to confirm the information arising from the qualitative data from the interviews.

Bi-polar scales were also used to facilitate in-depth exploration of particular areas systematically and objectively. The words used by the interviewees, obtained from content analysis of the interview data, were used for at least one pole of each scale. The use of scales allowed the respondents to express the degree and intensity of an attitude, and had the advantage of tapping sources of interest economically by giving maximum information quickly. The scales and structured questionnaires both gave results which were able to be analysed statistically in order that differences between individuals and groups could be compared with a high degree of objectivity.

In spite of the richness of the qualitative and quantitative data obtained from such techniques, the use of more formal psychometric test methods was also employed. Such tests yield data in a uniform, structured, valid and reliable manner which are able to be analysed accordingly. Used in conjunction with methods yielding qualitative data one confirms and substantiates the other. The tests selected, Cattell's *Sixteen Personality Factor Questionnaire* and the *Embedded Figures Test*, both have well established norms from which relatively reliable comparisons could be made.

Figure 3.1 gives an overview of the strucure and measures used during the research.

The questionnaires

The first major questionnaire, the *General Questionnaire* asked for demographic data and some basic details about the computer dependent respondents' computing activities. The main aim was to establish whether these respondents represented a cross-section of the general population. The results indicated that they did not and an extensive survey, the *Schools Survey*, was undertaken to discover why other groups were not represented. This involved interviewing teachers of computer studies and the administration of a questionnaire to secondary school children.

Ten of the computer dependent respondents were then interviewed in order that their responses could guide the other phases of the research. These interviews were based upon the interview schedules used with the undergraduate students, but with new sections being added to cover the areas of relationships, marriage and work, and with the computing sections more detailed and specific. From the data obtained, the main interview schedule was further developed, refined and piloted; the standardized psychometric tests were selected; and the questionnaires and scales were developed.

The remaining measures which were specifically devised for this research are summarized below:

- The *Computing Questionnaire* examined computer programming, hardware and networking interests (see Chapter 6);
- The *Leisure Activities Inventory* isolated the past hobbies of interest to the respondents to examine relationships between them and their computing activities (see Chapter 7);
- The *Attitudes to Computers Scale* and *Attitudes to People Scale* examined

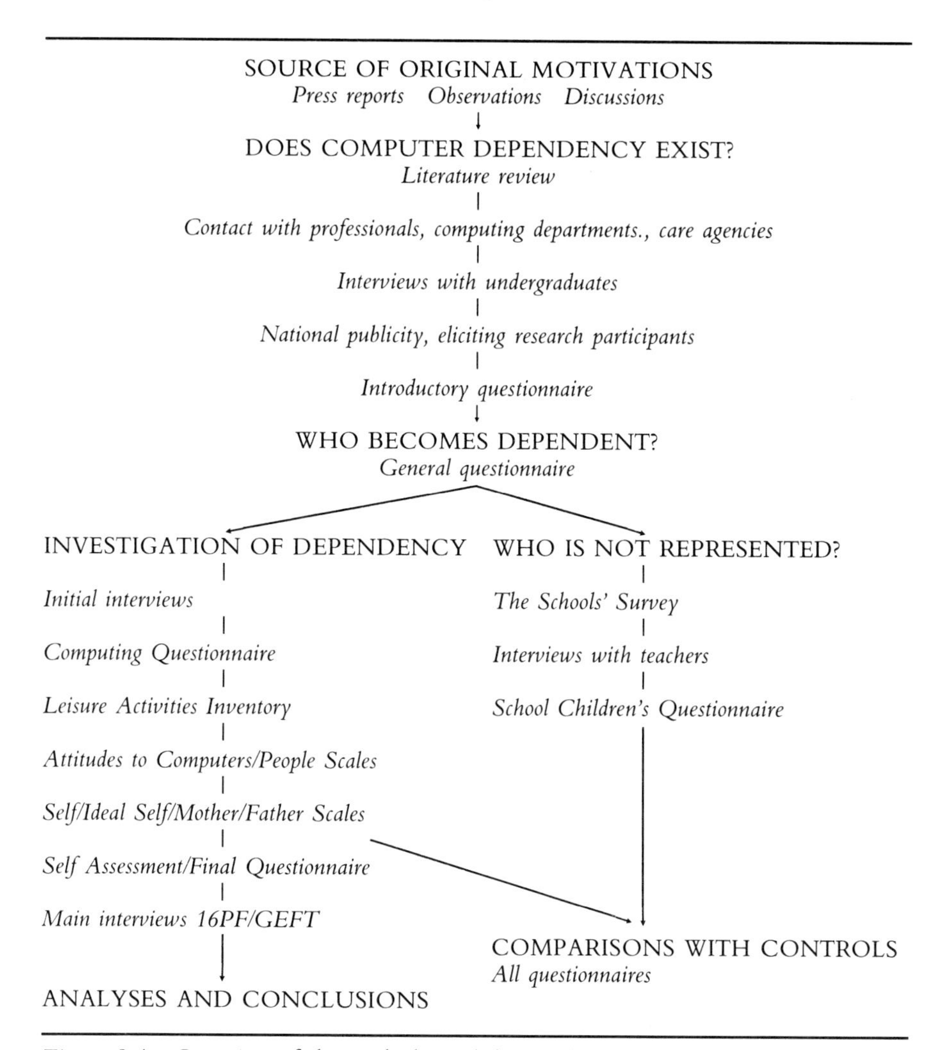

Figure 3.1 Overview of the methods used during the research project

whether the respondents viewed computers more favourably than humans (as was suggested during the first ten interviews, see Chapter 8);

- The *Attitudes to Mother Scale* and *Attitudes to Father Scale* examined whether negative reports from the first interviewees about their parents were confirmed by the rest of the group (see Chapter 8);
- The *Self Report Scale* and the *Ideal Self Scale* were designed to measure differences between the rating of their own and their ideal personalities (see chapter 8);
- The *Computing Self Assessment* and the *Partner's Assessment* asked the

respondents and their spouses to estimate proportions of time spent on computing activities (see Chapters 6 and 7);
- The *Final Questionnaire* summarized many of the details discussed during the interviews, both concerning the respondents' personal lives and other aspects of their computing (see Chapters 4 to 10).

The data are divided within the thesis according to the subject area, although they were not necessarily all drawn from one source. For example, the discussion of the computing activities in Chapter 6 was drawn from data from the General Questionnaire, the Computing Questionnaire and the Final Questionnaire.

Standardized tests

The standardized tests which were selected to be administered during the interviews were the *Sixteen Personality Factor Questionnaire (16PF)* and the *Group Embedded Figures Test (GEFT)*. In total, 45 of the respondents were able to be interviewed and given the standardized tests, (the first ten were re-interviewed in order to ensure that their data were complete). All, whether interviewed or not, completed the same sets of questionnaires and scales.

Cattell's Sixteen Personality Factor Questionnaire (16PF) was selected as appropriate for this purpose; it has gained wide recognition and is one of the most commonly used tests available for the measurement of general personality. In spite of criticisms of the 16PF (see for example Buros, 1972 and Mitchell, 1983), it was considered adequate for the purpose intended. The results were not to be used in a clinical setting nor for vocational prediction, but to add further clarification to the existing data. The 16PF has comprehensive tables of British norms for both adult females and males, and as no children under 16 were interviewed all 45 Dependents were able to complete the same version of the 16PF; Form A (IPAT, 1979).

The majority also completed the *Group Embedded Figures Test (GEFT)*, (Oltman *et al.*, 1971). This was included to examine whether the Dependents did exhibit the cognitive styles of field-independence and convergence as implied by many of the earlier results. (The group version of the Embedded Figures Test had been selected as it had been originally hoped that comparisons could be made with control subjects, whom it was thought could be assembled in one place to complete the test. This proved impossible. However the GEFT was still considered to be the most suitable version to use as it involves less interference from the tester and allows for fuller control and autonomy on the part of the testee.)

As with any research employing postal techniques, especially those which require follow-up questionnaires, there was the risk that many of the volunteers would be lost as the work progressed if they were not fully conversant with what was expected of them. The respondents were not being paid for their participation and only by maintaining their interest could their cooperation be assured.

Initial communications with the volunteers explicitly stated that their participation in the research could be expected to continue for some months, and that this would involve the completion of numerous questionnaires, together with an interview whenever possible. They were informed that they would be asked questions about themselves, their personalities and their backgrounds, as well as questions about their computing. The aim was to ensure that they were fully conversant with the depth of information required in order that they were not later surprised or offended by the questions which were considered relevant. They were also assured that any information they gave would remain confidential, with any quotations used anonymously.

Over the course of the research six separate postal communications were made, some incorporating more than one questionnaire or scale. Understandably, with each successive mailing slightly fewer people responded. Although only 75 of the 121 respondents (62 per cent), who first made contact and considered themselves to be computer dependent, completed and returned all six sets of questionnaires, the response rates between each mailing were extremely high. (Table 3.1 gives a breakdown of each mailing with the response rates for each.)

Table 3.1 Response rates from the six mailings to the Dependents

Mailings	Frequency	Percentage rate
1 Introductory Questionnaire	121	
2 General Questionnaire	106	87·6%
3 Computing Quest/Leisure Activities Inv.	96	90·6%
4 Computers/People Scales	83	86·4%
5 Self/Ideal Self/Mother/Father Scales	77	92·8%
6 Final Questionnaire/Self Assessment	75	97·4%

These response rates were higher than one might have anticipated for postal methods, but it is essential to remember that those who wrote did so voluntarily and were not selected through quota sampling or other methods. Their interest existed at the outset and it was the task of the communications which followed to maintain this interest, which it appeared to do very successfully. Any loss of respondents was disappointing; however as the length of time between the first and the last mailings was more than a year, the response rates appeared more than satisfactory.

The need for and elicitation of control groups

At the outset of this research the intention had been to allow the data from the dependent computer respondents to stand alone, as that from a cross-section of the general population. However, the research revealed that this pool of respondents was not a cross-section but showed particular demographic characteristics, therefore any results obtained could merely be artefacts of the group's composition rather than due to computer dependency. In order to isolate the factors pertinent to computer dependency satisfactorily it was found necessary to make comparisons between the computer dependent participants and their peers.

The computer dependent respondents differed from the general population on the demographic variables of age, sex distribution, and highest educational level obtained, and therefore it was upon these criteria that matched control groups were selected.

Two groups were selected with whom comparisons were made with the computer dependent respondents; a matched group of non-dependent computer owners and a matched group of non-owners of computers. The groups are labelled as the *Dependents*, the *Owners* and the *Non-owners* for clarity. It was found necessary to select two additional groups to act as controls based on the following assumptions:

(1) It was assumed that those who did own computers would differ in interests and attitudes from those who did not. Therefore comparisons made between the Dependents and a mixed group of people consisting of both owners and non-owners of computers, however well-matched on the demographic variables, would present confounding variables when differences were investigated.

(2) It was assumed that by 1984 the majority of those seriously interested in computers would own their own personal computers.

(3) It was assumed that only a small percentage of home computer owners would consider themselves to be computer dependent.

(4) It was assumed that if there were variables pertinent only to computer dependency these would be apparent if significant differences were found between the results of the Dependents and the two control groups, the Owners and the Non-owners.

(5) It was assumed that if there were variables which differentiated computer ownership from non-ownership these would be shown by any significant differences found between the results from the Non-owners and the two computer-owning groups, the Dependents and the Owners.

Following these assumptions it would therefore prove possible to isolate the characteristics pertinent to computer dependency and not merely to computer ownership by making comparisons between the three matched groups.

Seventy-five Dependents completed all of the questionnaires and inventories, and these people formed the group with whom comparisons were made so that complete sets of data could be analyzed. (Of the

Dependents who did not complete all the measures, analyses revealed that they did not differ demographically from those 75 who had completed all the tasks, therefore no undue bias was created by not using their data.)

Each individual within the Dependent group was matched on the criteria of age, sex and highest educational level with one Owner and one Non-owner with the same characteristics in order that the same spread of variability be represented within each group. (Because of the wide spread of ages represented within the Dependent group they were divided into bands of five-year age groups to facilitate this task of matching.) Table 3.2 shows the breakdown of the 75 respondents by sex, age band and highest educational level with whom the control groups were matched.

Table 3.2 Age and educational level of the 75 Dependents who were matched with controls

Educational level	1	2	3	4	5	6	
Age bands							Total
10–14	2						2
15–19	1	5	6				12
20–24	1	1	2	3	5		12
25–29			1	3	2		6
30–34	1(1)	2(1)	3(1)	2	6		14
35–39	1	3	2	2	6	2	16
40–44				1	2	4	7
45–49						1	1
50–54					2(2)		2
55–59					2(1)		2
60+						1	1
Total	6	11	14	11	25	8	75

Key: Educational level
1–No formal qualifications
2–'O' level or equivalent (16 years)
3–'A' level or equivalent (18 years)
4–Technical qualification
5–First degree or equivalent
6–Higher degree or equivalent
1–number of females

The control group volunteers were told that the research was to investigate the personalities, backgrounds, education and interests of people who owned or did not own computers. Computer dependency was not mentioned and there existed therefore the possibility that a few of the control group of computer owners were themselves also dependent upon computers. However, as so few of the computer-owning population were thought to be dependent it was felt that this factor would not greatly interfere with the results, especially as any resulting differences found between the groups would as a consequence prove more significant because of their presence.

Individual matching ensured that the groups would show the same distribution of demographic variables for which they had been matched, and this ensured that to a very large extent like was being compared with like. It should be noted that although the respondents were matched individually with the controls all the resultant data were analysed by group rather than individual, as it was the gross differences between the three groups which were of interest.

The statistical methods used

Although the initial investigation had suggested some characteristics of computer dependency, there was little knowledge as to exactly how the respondents would differ from the control groups carefully matched with them. Without wishing to prejudice the data by suggesting *a priori* trends, a device was needed which would allow the data to specify the trends and differences. Omnibus, devised by Meddis in the late 1970s, proved ideal for the demands of this research. Its mathematical basis is fully explained in Meddis (1980) and its statistical theory in Meddis (1984). It is a computerized package, using an analysis of variance by ranks procedure on non-parametric data, which gives one the ability to test hypotheses at any stage of the procedure. The test incorporates a group of non-parametric tests into one package, which capitalizes upon the common principles of them all. Meddis stresses that it is not a suite of tests, as one would tend to find in other statistical packages, but a single computational algorithm which is mathematically equivalent to rank sum analysis and gives exactly the same statistical results. Omnibus was particularly useful for the analyses of these results as the test will accept raw scores, ranks or frequency tables. It converts the data into ranks (if necessary), gives the sample rank means, tests their variability, and from this the probability and the significance levels are specified.

The first analysis carried out upon the data always assumed the Null hypothesis, that there was no significant difference between the three sets of data and that they were drawn from the same population. The first test was therefore non-specific, but if this hypothesis proved false a specific test could be carried out immediately.

For the purpose of this research it was assumed that if differences occurred between the groups these differences would be greatest between the Dependent computer owners and the Non-owners, with the results from the non-dependent Owners lying somewhere between or tied to one of the other groups. If a variable was particularly pertinent to computing the results from both groups of computer owners would differ significantly from those of the Non-owners, and if only relevant to computer dependency the results from the Dependents would differ from those of the two control groups. The provision of the sample rank means enabled the choice of trends and contrasts to be made easily. Omnibus proved invaluable for this type of exploratory research where hypotheses can develop from the data presented.

It was considered that the use of carefully matched control groups would provide an efficient method to enable the pertinent variables to be isolated with a high degree of accuracy. The aim was to produce data which were clear, meaningful and precise, and in many respects the simpler the statistic the more it can fulfil these objectives unambiguously. The use of Omnibus, designed for the analysis of non-parametric data, proved very adequate for the demands of this research.

Chapter 4
Analysis of the demographic data: Who is computer dependent?

The demographic characteristics of the computer dependent respondents

The first area of investigation was to determine the types of people who considered themselves to be dependent upon computers. The demographic variables were investigated via a General Questionnaire, and the results showed that the respondents did not form a cross-section of the population.

The data revealed that 100 of the 106 respondents were male, with a mean age for the total group of 29·7 years ranging from 14 to 64 years of age. Details of their marital status are summarized in Table 4.1. Of the 45 people who were or had been married 35 had children, as did one single person. All of the unmarried respondents were male, 15 of whom were over the age of 30.

Table 4.1 Marital status of the Dependents by age group and sex. The number of females within a group is given in parentheses.

Age range	Single	Married	Separated	Divorced	Widowed	Totals
10–19	22					22
20–29	24	6				30
30–39	13	24(4)	1	1		39
40–49	1	8		1		10
50–59	1	1(1)		1(1)		3
60+		1			1	2
Totals	61	40	1	3	1	106

Details of their highest educational level achieved revealed that of the 84 respondents over the age of 20 years, 68 (80 per cent) had at least gained 'A' (advanced) level examinations (taken at age 18) or their equivalent, of whom 18 (15 per cent of the total) had higher technical qualifications, and 37 (31 per cent) held degrees. Table 4.2 details their highest educational level by age group.

Table 4.2 Educational levels achieved by the 106 Dependents who completed the General Questionnaire, by age group

Age	1 None	2 'O' level (age 16)	3 'A' level (age 18)	4 Technical	5 1st degree	6 2nd degree	N
10–19	5	10	6	1			22
20–29	2	3	4	12	9		30
30–39	2	6	10	4	13	4	39
40–49	1	1		1	2	5	10
50–50					3		3
60+	1					1	2
Totals	11	20	20	18	27	10	106

Preliminary observations

The responses to the demographic questions yielded some unanticipated results, the most significant being the sex distribution of the subjects. This was very highly biased towards males who formed 94 per cent of the sample.

No women had been suggested as being dependent upon computers when the undergraduates were being sought for interview, although this may simply have been due to the fact that the Loughborough University student population was predominantly male, in the ratio of approximately 2 : 1. Re-examination of the literature revealed that if the sex of a computer dependent person had been mentioned, that person was always male. This may be due to the fact that many authors use the term 'he' generically, without specifically meaning to refer to males alone, but Tittscher (1984), in a review of various retail surveys, discovered that 96 per cent of home computers were purchased by males, and that 88 per cent of the main users were male. It seemed more likely that this result accurately reflected the relevant proportions of the sexes within the population of computer owners, of whom the Dependents may be a cross-section. This point will be addressed further in Chapter 5.

It had been anticipated from press reports and previous literature that the most likely ages for people to be dependent upon computers would be those covering the school and university years, although this again could merely reflect the situations of the observing authors. The result, giving a mean age of nearly 30 years for the Dependents, was therefore higher than had been anticipated. However, a survey conducted by Mintel (1984) confirmed that the age range obtained did in fact accurately reflect the incidence of computer ownership in the UK. Dividing the respondents into the age bands used by Mintel, analysis showed there to be no significant difference between the two sets of data (see Table 4.3).

If the statements in the literature were to be relied upon, the marital status of these respondents was also exceptional. In his book *Silicon shock:*

Table 4.3 Frequencies of computer owners (Mintel, 1984) and Dependents (observed, 1984), by age groupl. (p>0·10)

Age band years	Computer owners N=159	Dependents N=101
15–19	35	18
20–24	22	15
25–34	31	34
35–44	42	28
45–54	22	2
55–64	7	4
65+	0	0

the menace of the computer invasion, Simons (1985) devotes one complete section to marriages under threat from the intervention of the computer into relationships, giving examples of very high divorce rates within the 'Silicon Valley' area of California. *The Sunday Express* (29 January, 1984) also reported work from a marriage counsellor in California who, it was said, studied 3000 couples and discovered that:

> 'Nearly 50 per cent of those couples got divorced and most of the divorces can be directly attributed to a "love triangle" with the computer.'

If computing caused such problems within British marriages, one would expect similarly high divorce rates to appear within this sample of people. This was not the case. From the data personally requested from the Office of Population Censuses and Surveys (OPCS), covering the 1981 census of Great Britain, it was possible to analyse that 3·9 per cent of the population between the ages of 20 and 64 were divorced. The results for the same age range for this group of Dependents showed a divorce rate of 3·6 per cent; an almost identical percentage. Even without formal analysis it was possible to observe that, within this British sample at least, computer dependency did not appear to cause problems within relationships serious enough to warrant the total breakdown of marriage. One is led to assume that there are cultural differences between the United Kingdom and the United States concerning divorce.

A seemingly large percentage of the males over the age of 30 years were single, 27·8 per cent. Using the data from the OPCS for the 1981 Census, it was possible to discover that 10·7 per cent of the male population between the ages of 30 and 64 were single; consequently there was a much higher rate of single men within this group than might be expected. One could argue that these results reflected the comments from the literature, that computer dependent persons have a higher incidence of social problems which could perhaps result in a lower rate of marriage. However, as such a large proportion of this sample had undertaken tertiary education it is perhaps natural to assume that those who continue their education for a number of years beyond school-leaving age would be expected to delay marriage longer than those without such qualifications.

The very high level of educational qualifications among the respondents had not been anticipated, despite the fact that most of the literature cited university students and graduates becoming dependent upon computers. It had been assumed that this occurred because the observations had been made by academics within university environments, who had merely observed those closest to them. It was expected that the respondents would represent the full gamut of educational achievement seen in the general population. However, these data seemed to confirm that the computer dependent person is often highly qualified, and that the literature and interviewed students quite accurately reflected the computer dependent population, if only accidentally. Data from the 1981 British Census showed that only 10 per cent of the general population gained 'A' levels and above (the percentage for the Dependents was 80) and only 5 per cent had degrees (the percentage for this sample was 31).

One could deduce from these data that only well-qualified individuals volunteer to take part in research projects, that these people are more likely to become dependent upon computers, or that in general it is those with higher educational qualifications who tend to become computer owners, with the Dependents forming a cross-section of that population. No literature could be found to confirm these suppositions, although the Mintel survey (1984) discovered that those in the A/B social groups formed the highest proportion of computer owners. If it is believed that the A/Bs are more likely to have undertaken tertiary education, then these results also tend to confirm that the Dependents formed a cross-section of the population of computer owners.

One could also argue that those who have experienced higher levels of education are more likely to find a need for the computational powers of a home computer, are more likely to have the finances necessary to purchase one, or are more likely to have had prior experience of computers during their education and work and are therefore more enthusiastic about new technology than other members of society. In addition, to use a microcomputer at all successfully one needs a certain amount of ingenuity and drive in order to discover its capabilities and capacities. This involves the study of literature and manuals, and a determination to succeed in spite of the poor quality of many of them. This in itself could act as a form of self-selection. One cannot say that high intelligence is a prerequisite for using a computer, nor are all intelligent people computer owners, but an inquisitive, persistent and enquiring mind could be seen as essential in order for the interest to develop to the extent of wishing to purchase and use a computer in one's spare time. The results seemed to confirm the statement by Large (1984):

> 'By the spring of 1984 more than 10% of British homes had a computer and market surveys suggested that UK sales would continue at a rate of around a million a year for at least another year or two. Although the rush

meant that Britain had more home computers per capita than any other country, the advance in 'computer literacy' was common across the west. The worrying fact was the inevitable one, that the advance was also predominantly among the already-alerted middle classes.'

Although no firm conclusions could be drawn, it was apparent that the respondents who considered themselves to be dependent upon computers did not form a cross-section of the general population. The majority were male, had received high levels of education, and yielded an average age of 30 years. There was no logical reason to suppose that the publicity used to attract respondents was biased to this section of the population, nor that only the well-qualified male would have read and responded to the articles, therefore it could be assumed that the observed proportion of males did in fact truly represent the computer dependent population. Nevertheless any data obtained from these respondents, however unusual it may have appeared, may not have been found to differ from that obtained from any other similar groups of males; hence the need to compare the Dependents with matched control groups in order to isolate the features particularly relevant to computer dependency.

Demographic comparisons with the control groups

The two control groups had been matched with the Dependents on the variables of age, sex and highest educational level obtained; the three factors which demonstrated that the Dependents were not drawn from a cross-section of the general population. However, further demographic characteristics of the groups were also investigated, as the results from the questionnaires and the initial interviews with the Dependents suggested that other factors may have been associated with the development of computer dependency, as follows:

- Fewer Dependents were married than within the general population;
- More of their marriages than perhaps expected were childless;
- Some of the Dependents considered their parents to have been older than typical, leading to a more conservative, restricted life-style;
- The early death of a parent when the respondent was young also seemed to be a major influence in the shaping of some personalities;
- A high proportion of the Dependents were first born children (over 50 per cent);
- The majority of the Dependents and their fathers had occupations within the areas of science and technology.

As the Dependents seemed to differ from the general population in these respects, investigations were carried out to examine whether these factors also differentiated them from their peers; the matched control groups of Owners and Non-owners. The data were obtained from questions included in the General Questionnaire and the Final Questionnaire, and the results from the analyses of the raw data showed that there were in fact further demographic similarities between the three groups of respondents.

Although a larger proportion of the Dependents had been shown to be unmarried than was found in the general population, comparisons between the Dependents and the controls showed no significant difference in their marital status (Table 4.4). The groups' combined results showed 52 per cent single, 41 per cent married and seven per cent separated, divorced or widowed. One can therefore assume that many of those who continue their education beyond the secondary school level are likely to marry later, or perhaps not at all. The Dependent group were also no more nor less likely to have children than the other groups, reflecting further similarities in lifestyle. More importantly, the rate of divorce and separation within the Dependent group was not significantly different from that in either of the control groups. One may therefore conclude that computer dependency is no more likely to cause marital breakdown than other factors. In this respect the marriages of British computer dependents appear to differ from those of their American counterparts of whom, it is claimed, many are divorcing because of this syndrome. (The problems which were caused with the Dependents' marriages are discussed in Chapter 10).

Table 4.4 Marital status of the three groups. ($p > 0 \cdot 10$)

Marital status	Dependents	Owners	Non-owners
Single	37	34	46
Married	33	36	23
Separated	1	2	1
Divorced	4	3	5

Although the early death of a parent had been interpreted as relevant to the shaping of personalities and attitudes of a few of the Dependent interviewees, this event was shown to be no more likely to have occurred to the Dependent group than to the Owners or Non-owners. Approximately 10 per cent of the respondents from each of the groups reported the early death of a parent, that is before the respondent had reached the age of 18 years old. This does not suggest that such a trauma is insignificant in a child's life, but demonstrates that one group was no more likely to have suffered this trauma than another.

The seemingly high proportion of first-born children amongst Dependents was also shown to be typical of the matched control groups (a grand mean of 52 per cent), as were the reported ages of parents at the respondents' birth. The results showed the combined mean age of fathers to be 31·6 years and mothers to be 28·8 years when the respondents were born; perhaps a little older than expected especially as so many of the respondents were first-born children. However, as the occupations of the respondents' parents suggested that many of them had received tertiary levels of education, this could explain why many had deferred marriage and childbearing.

The occupational data did highlight the one significant difference which was observed between the three groups of respondents, however. Although the majority of all three groups were in occupations which demanded tertiary level qualifications, both the Dependents and the Owners, together with their fathers, were shown to be significantly more likely to work in the areas of science and technology than were the Non-owners and their fathers. This difference suggests that an interest in computers is directly linked with earlier influences and interests, factors which were further investigated and are discussed in Chapter 7.

One may conclude that the majority of these results indicate that all three groups of respondents were drawn from the same population (as summarized in Table 4.5). Any subsequent difference found between them can therefore be seen to be due to factors other than those of a demographic nature. The results also demonstrate the need for controlled comparative methods in exploratory studies such as this. By allowing the Dependents' data to stand alone, or by making comparisons between them and the general population, invalid conclusions could have been drawn. The Dependents were shown to be no more likely to be single, first-born, or divorced than their peers, for example. The observed similarity of the demographic characteristics of the groups therefore increases the likelihood that any further significant results are a facet of the main characteristics of the three groups; namely computer dependency, ownership or non-ownership of computers.

Table 4.5 Comparison of demographic variables between three groups

Demographic variable	Significance levels
Age of participant	(controlled)
Sex	(controlled)
Highest educational level	(controlled)
Marital status	n/s
Number of sons	n/s
Number of daughters	n/s
Birth order in family	n/s
Age of father at respondent's birth	n/s
Age of mother at respondent's birth	n/s
Early death of parent	n/s

Chapter 5
Sex differences in computer use and dependency: Why were there so few females within the sample?

Introduction

It was apparent from the demographic data that the majority of the respondents describing themselves as computer dependent were male, in the ratio of 100:6. This sex distribution was the most significant of all the demographic results and was therefore considered worthy of study in more depth.

The comprehensive publicity for this research was not biased to exclude women, nor was there reason to suppose that women would be less likely to respond than men. One could therefore assume that the proportion of women who responded accurately reflected the number of female computer dependent people within the population. If this were so, why are females less likely to become dependent upon computers? Are they generally less interested in computers, or if they are interested do they use computers in such a way that dependency is inhibited?

It is not unreasonable to surmise that more adult males than females have had the opportunity to interact with computers while at work. Nor is it unreasonable to assume that the majority of those women who do work with new technology tend to be concentrated in the areas of data-entry and word-processing, while men are more often found in the creative worlds of programming and system analysis. This lack of opportunity to use computers as inventive, ingenious tools could inhibit women from viewing computers as machines for leisure and creativity, and both the pre-defined tasks and lack of exposure could explain why so few women appear to exhibit the symptoms of computer dependency. However, there may be distinct attitude differences towards computers between the sexes which could have little to do with either the task or the exposure. This section of the research aimed to investigate these aspects. By controlling exposure to computers and the type of task being performed differences in attitudes between the sexes could be more easily isolated and investigated.

The survey to investigate attitude differences between the sexes was carried out with secondary school children. In spite of the fact that the age profile differed from those of the computer dependent respondents, both computer exposure and task function would be more constant between the sexes in the school situation than in any adult working sphere.

It was assumed that by late 1984, when this survey was initiated, the vast majority of children would have received at least some experience using microcomputers within their schools (following various governmental initiatives which aimed to ensure this). If mere lack of exposure to new technology caused females to be reluctant to use computers, any difference in attitude between the sexes should surely have diminished, if not totally disappeared, within the school-age group.

As it was assumed that both boys and girls would be using computers for similar tasks while in the controlled environment of a school, there seemed no reason to suppose that there would be differences in interest levels between the sexes. Both sexes should therefore show the same range of interest levels with the same likelihood that they become dependent users: unless of course something more profound was happening.

The schools' survey

The survey was conducted within the local secondary schools of the Leicestershire public sector. There was no reason to suppose that the county of Leicestershire did not represent a good mix of what may be described as 'affluent' and 'deprived' areas as well as rural and urban schools. The use of local schools also gave greater opportunity to carry out the investigations with efficiency and economy. The survey aimed to discover whether there were differences in attitudes, interests and uses of computers between boys and girls, and to discover if the syndrome of computer dependency was observed within school children whether it affected both sexes equally.

The opinions of computer studies teachers

To gain an overview of computing within schools, and before children were approached directly, the opinions of teachers were investigated via brief questionnaires, followed by interviews carried out on school premises. The questionnaire was sent to the main computer studies teacher within each of the 85 local state-funded secondary schools, and simply asked whether teachers had seen signs of computer dependency in any of the children they taught and if they would be willing to be interviewed about the use of computers within their schools. The interviews which were subsequently carried out were semi-structured to cover the following areas:

- Computer dependency among school children;
- Children's attitudes and use of computers in the classroom and in computer clubs;
- School computing—how it was taught, the examinations taken and the computers used by other teachers;
- Computers—their number, acquisition, storage and mobility;
- Teachers' training, background and sex.

Replies were received from 41 teachers, 19 of whom had seen signs of computer dependency within their pupils, and 23 interviews followed.

Of the 23 schools visited, 11 teachers believed that they taught children who exhibited the characteristic symptoms of computer dependency, and between them reported 26 boys and one girl. All of these children were computer owners. An analysis between these frequencies and those from the research group, of whom 100 of the 106 respondents were male, showed that there were no significant differences in the sex distributions of the groups, ($p > 0 \cdot 10$). It could therefore be considered that the sex distribution of both groups was representative of the whole computer dependent population, although there were no further means of establishing this.

The total number of children within these 23 schools was 19 520, giving the proportion of children considered to be dependent upon computers by their teachers to be approximately 1 in 700 (0·14 per cent). Later results showed that of the children who completed the School Children's Questionnaire 44 per cent were computer owners. If these percentages were representative of all the children within these 23 schools then one in 300 computer owners were described as dependent (0·33 per cent).

There were many similarities between the teachers' descriptions of the children thought to be computer dependent. Invariably these children were described as very intelligent, and 22 of the 27 children were considered to be socially inadequate in some ways, with some coming from unhappy home situations. Most of these pupils were considered to have experienced social difficulties before using computers, and some teachers believed that computing had been of benefit in developing their self-confidence and increasing their prestige; comments not unlike those from the undergraduates when describing themselves during interviews. Not all teachers were as positive, however, and some felt that the pupils' use of computers was increasing distance between them and their peers. Some also considered that these children's dedication to computing could damage their academic careers. Even if the children were thought to be able to achieve the grades necessary for entry into tertiary education, some of the teachers believed that they would not be accepted because of their poor social skills and lack of interest in other subjects.

Explanations from the teachers for the greater likelihood of boys becoming computer dependent were various. They felt that boys tended to treat the computer as a toy rather than as a problem-solving tool, and to hold a generally more immature attitude towards their work. They were also considered to be more likely than girls to have just one serious hobby, and the ability to excel at one activity was considered more important to boys who were thought to be naturally more competitive. These facets were felt to lead some boys with poor social skills to become 'obsessive' about computing. Computing skills tended to be admired by other children, even if not overtly, and the safe interaction with the computer was felt to form a large part of the computer's appeal to boys.

All of the teachers observed that not only were boys more likely to become dependent upon computers, but in the main it was the boys who were more interested in new technology. This was reflected by the numbers of children taking computer studies examination courses in the school year 1984–5. Of the 1836 children taking the courses in these 23 schools, 1309 (71·3 per cent) were boys. National statistics (Department of Education and Science, 1985) showed that 71·9 per cent of the children entered for GCE* computer studies/science examinations in the summer of 1985 were boys. One may conclude that the results from these Leicestershire schools seemed to represent those observed nationally. Only when computer studies courses were compulsory for all members of a class were the sex distributions normal, and this occurred in only ten of the schools visited in which computer awareness courses were offered.

These results seem to confirm that girls were significantly less interested in computers than boys, in spite of the teachers' assurances that both sexes were given equal opportunity to use computers and did use them for the same type of tasks. The teachers put forward many theories and explanations for the sex differences, which are summarized below. Some of these were substantiated by the literature, and although not all of the references found refer directly to computing they are relevant to the wider educational issues raised.

(1) Computing was linked to mathematics and science subjects within schools, subjects which have been traditionally viewed as 'masculine', and as a consequence female prejudices and lack of confidence in these subjects were transferred directly to computer studies. (Of the 255 computers in these 23 schools, 213 were sited in one of these three departments). Computer studies teachers were invariably male, in the ratio of 3:1, reinforcing this attitude in all children. (Committee of Enquiry into the Teaching of Mathematics in Schools, 1982; Equal Opportunities

---

**FOOTNOTE* General Certificate of Education taken at 16 years of age.

Commission, 1983; Saunders, 1975; Weiner, 1980; Ward, 1983 and 1985).

(2) The siting of the computers discouraged their use by other teachers even if they wished to be involved. Security needs often demanded that all the equipment be confined to one room which could be securely fastened, preventing them from being moved easily to other areas of the school.

(3) Female teachers were often as disaffected by new technology as their students. They were less likely to own a home computer and therefore less likely to be familiar enough with new technology to wish to attempt to use them in the stressful situation of the classroom, especially in front of experienced and knowledgeable boys.

(4) Inadequate training in the use of computers for teachers from other departments ensured that computers remained in the domain of computer studies.

(5) The need to concentrate upon national examinations in computer studies made the equipment unavailable for other tasks.

(6) Boys enjoyed programming and learning about the hardware, whereas girls wanted to use the computer as a tool without wishing to learn to program and did not see the computer as intrinsically interesting. Computer studies was not seen as relevant to the girls' lives and they appeared to demand more practical, functional end-products from computing than were possible within the school environment. On the other hand, the boys' intrinsic interest in computers was able to be satisfied. (Equal Opportunities Commission, 1983; Forrest, 1984; Rauta and Hunt, 1972; Ward, 1983).

(7) The content of the national computing examinations was more closely related to boys' rather than girls' interests.

(8) Boys tended to be confident in their dealings with computers. They rarely planned their programs adequately but enjoyed the necessary debugging which resulted. Girls tended to plan carefully, perhaps through fear of using the computer, and expected the program to run at the first attempt. Boys rose to the challenge of debugging, girls took it as a personal affront and often felt a sense of failure.

(9) Some teachers felt that computer studies courses were irrelevant to the children's need, especially the concentration on BASIC programming (an opinion echoed by Nash *et al.*, 1985), but attempts to design more practical courses and forego national syllabi were met with resistance from parents who demanded external qualifications for their children.

(10) Girls tended to opt out of 'masculine' subjects which they found difficult (Keys and Ormerod, 1976), and because so few girls chose to take computer studies this option was often pitted against such topics as home economics, languages and biology, the traditional 'feminine' subjects, thereby perpetuating this bias.

(11) The few girls who did opt to study computing were often unable to achieve more than the minimum marks, in spite of their careful program planning. The teachers thought girls were less likely to have a home computer, without which they could not keep up with the mass of practical work demanded by the examinations as there was insufficient equipment and time available at school.

(12) Touch typing, which was said to be of interest to many girls, was not taught on computers in the schools visited (but confined to typewriters in female-dominated rooms). Word processing packages were also rarely available. This prevented the children from using computers as tools for creative writing, a facet said to be of interest to more girls than boys.

(13) Funding was inadequate to provide educational software to run on the computers, therefore they could rarely be used as tools without first writing a relevant program.

(14) The software which was available was sometimes educationally unsound, being written by programmers rather than teachers, and was designed to take advantage of the properties of the computer rather than the needs of the child (much tended to be suitable for rote-learning and multi-choice questioning only).

(15) Because of the lack of software, teachers were often forced to write their own programs, and those who were experienced enough to do so tended to be the male teachers writing programs for science, mathematics or computer studies, thereby perpetuating the existing biases.

(16) Boys had more experience in using new technology outside school and were therefore less inhibited when using the school equipment and tended to dominate it. (Kelly, 1982; Ward, 1985).

(17) Poor equipment reliability increased girls' intimidation, and boys tended to dominate the better machines because of their greater awareness and understanding of the hardware.

(18) Boys demanded more teacher attention which could not be ignored if discipline was to be maintained, therefore their needs were met more fully, (Elliot, 1974; Mahoney, 1985; Stanworth, 1981). Although not mentioned by the teachers, Clarricoates (1978) found that both boys and girls recognized this 'hidden curriculum', and girls became very reluctant to ask for assistance, believing themselves to be second class citizens within the classroom.

(19) Boys believed computers were easy to use, whilst girls believed they had little ability. Failure and the need to debug programs reinforced this belief in girls leading to a form of learned helplessness and poor achievement motivation, (Dweck *et al.*, 1978; Fennema, 1981).

(20) Computers invariably had to be shared in classes and if the groups were of both sexes the boys used the keyboards while the girls watched. Some teachers found that they had to separate the sexes if the girls were to have hands-on experience (as recognized by Mahoney, 1985).

(21) Although the attitudes of the boys and girls towards computers could be assumed to develop at puberty, when sex-role stereotyping becomes highly pronounced, they were also found to exist in primary schools (Russell, 1986).

(22) Even in the unrestricted atmosphere of the schools' lunchtime computer clubs there was little of interest for the girls to do. Most of the available recreational software consisted of computer games of the 'space-invader' type which were male oriented, and the lack of close supervision ensured that boys dominated the equipment and intimidated the girls.

(23) Because of the girls' general disinterest and disaffection with computing, even if they were given exclusive computer club time or were provided with software which was more in keeping with their interests, they were still unlikely to choose to use computers.

(24) The teachers did not believe that girls were inherently unable to use computers successfully, but attributed it to lack of relevance, sex-role stereotyping, the masculinization of the subject, and poor achievement motivation. (McClelland *et al.*, 1953; Hunt, 1965; Horner, 1970; and Atkinson and Raynor, 1974). This was demonstrated by the fact that two groups of girls were successful and did enjoy computing; those of Asian origin and those attending a single-sex school.

Discussion of the teachers' opinions

Although I had observed disaffection expressed towards computers by many adult females, within and outside the university environment, it had been anticipated that the differences between the sexes in computer use and computer dependency would have been significantly less within the younger generation of school children, but the results indicated that this was far from so. The teachers believed that distinct attitude differences still existed between the sexes even for adolescents who had grown up with computers in their schools and homes. Although some of the differences were seen to be caused by school organization, teaching methods or availability of resources, many of the attitudes appeared to be more deep-seated.

One of the most significant of the findings from the interviews was that computing was almost universally considered by both sexes to be a masculine activity which males found easy, and the more closely a subject seemed to appeal to male modes of thinking and working the more alienated females appeared to become. The females appeared to be more oriented towards practical uses of the computer and were said rarely to be interested in their intrinsic merits. Not only did the school situation inhibit the use of computers for practical purposes, because of the restrictions of the examination curricula and lack of software, but so too did the use of a home computer at the time of these interviews. In 1984-5 word-processing software, for example, was not freely available for microcomputers and was very expensive, while most cheap software tended to be in the form of games. In addition most home owners did not own printers, which at this time were often more expensive than microcomputers, thereby rendering word-processing software almost useless. Without a desire to play games or to program there was little of a practical nature which could be carried out on a small computer.

The teachers also suggested that in addition to boys being more interested in the intrinsic merits of computers and learning programming languages, the end-product could sometimes be seen as a secondary by-product of their work rather than the primary aim. If one uses the analogy that the computer is a tool used in school in much the same way as a textbook, the boys seemed more prepared to write, typeset and also print the book, while the girls wished to be able to use a comprehensive index to enable them to find the relevant chapter so that their questions could be answered. The difference of course is that although it is not felt necessary to teach bookbinding in school, there are specific courses designed to teach children the intricacies of computer languages and programming. That many of the teachers of computer studies found these classes unnecessary and irrelevant for the majority of their pupils became very apparent during the interviews. Many wished to concentrate on the applications and implications of new technology, but the demands of the current examination system, parental pressure and lack of adequate software ensured that the current bias prevailed. Imbalances in the educational system allow such biases to continue. While girls are allowed to opt out of subjects which they find difficult or consider to be 'masculine', boys are frequently offered remedial help for such 'feminine' subjects as reading and writing; an imbalance seemingly condoned both by educationalists and parents, which suggests that the education of boys is still considered to be more important than that of girls.

Although the teachers concluded that computing was taught in secondary schools in a manner found to attract boys and alienate girls, there were of course exceptions to this rule; some boys were disaffected by new technology and some girls were very enthusiastic. Two groups of girls in particular were said by their teachers to enjoy computing more than most; those from the single-sex secondary school visited, and the girls of Asian origin. Not all the pupils from the girls' school were equally enthusiastic but their teacher (male), who had also taught at a mixed school, believed that the girls' negative attitudes were less apparent as they no longer had to compete with boys in the classroom. Although he had had to adapt his curriculum somewhat to accommodate the girls' interests, he found them to be far more interested in new technology than girls within mixed schools. Similarly, girls of Asian origin were reported to be as equally enthusiastic about computer studies as the boys in mixed schools, and were more likely to attend the computer clubs than other girls. Their teachers believed that Asian families held very positive attitudes towards all education, were prepared to invest in home computers for their daughters as well as their sons, and did not differentiate between the sexes in their expectations of success.

One may conclude from these results that the girls' disaffection with computers was partly culturally and socially induced, although all teachers continued to stress that the basic interests and demands from computers still differed significantly between the sexes. The British system of teaching computer studies to schoolchildren was therefore seen to discriminate between the sexes, and teachers saw little hope of change while the interaction with the computer remained so alien to the needs of the female.

Mere exposure to and experience with computers while young had therefore done little to remove the sex biases seen in the adult population; there was little doubt that schoolgirls held similar levels of suspicion and disaffection towards computers to adult females. Much research has been carried out to try to determine the causes of the negative attitudes towards computers held by many naïve users (see Oberg, 1966; Eason, 1976, 1977; Thimbleby, 1978; Brockner, 1979; Stewart, 1981; and Shore, 1985). Although most of this research has not centred upon sex differences *per se*, it has tended to concentrate upon clerical workers who transfer from typewriters to computers and therefore does deal with the attitudes and difficulties experienced by females. The authors speak of lack of control, confusion, anxiety, lessening of self-esteem and helplessness; all factors which the teachers attributed to the female pupils, and which appeared to be quite deeply entrenched due to the type of interaction necessary when dealing with most computer systems.

The results from the interviews with the school teachers suggested that there were distinctly different types of attitudes expressed towards computers, from those who were totally disillusioned (mainly female) to those who were enthusiastic, even dependent (mainly male). Numerous authors have also isolated different types of computer users according to the attitudes engendered by experiences with new technology. Most of these mirror the gross sex differences observed within this school survey, and may apply to the adult population and account for the fact that so few females responded to the publicity.

Poulson (1984) divided users into machine-oriented and machine-indifferent personalities, Brod (1984) into techno-centred or techno-anxious types, and Malde (1978) into efficiency-centred or person-centred users. Shore (1985) offered the binary-coded Types 0 and 1; with Type 0 the person who likes machines and puzzles and would rather play with computers than use them, and Type 1 who is primarily interested in using the machine for a specific, practical purpose but is hindered from doing so as computers tend to be designed by Type 0 people.

Gray (1984) divided computer users into four types along two orthogonal continuums; microphobic to microphilic, and gregarious to isolated. The gregarious microphobics he described as the Luddites, the phobic and isolated as those fearful of computers because of perceived difficulties or health scares, the gregarious microphilic as one who is an enthusiast and enjoys sharing his computer experiences with others, and the isolated microphilic was labelled by him as the 'hacker' or 'terminal junkie'. Brabant (1982) went even further and gives five quasi-religious types of computer users; the atheist who has no use for the computer, the pragmatist who sees it as a functional tool, the internuncio who sees the computer as functional but also interesting, the proselyte who explores the system, and the true believer to whom the study of the computer becomes an end in itself.

All of these 'types', whether isolated by supposition, observation or research, were deduced to be caused by attitudes formed either through experience or prejudice. Such attitudinal differences were shown still to be apparent within our schools, both among the children and staff, in spite of the common availability of computers within the classroom, with the groups invariably divided by sex.

Occasionally more positive attitudes among females have been observed where different methods and programming languages are used. Seymour Papert devised the computer language LOGO specifically in order to make the computer flexible enough for children to use with ease and understanding, in order that they could learn mathematics 'as a living language' (Papert, 1980). His methods followed Piagetian principles, allowing the children to develop their own learning patterns from concepts which have been broken down into their concrete, simple components. His aim was to ensure that the 'child programs the computer' rather than the child's thinking having to adapt to an alien language. Papert's wife, Sherry Turkle, carried out research in a private American primary school, purpose-built to house a 'computer culture' where children could use computers and LOGO freely (Turkle, 1984). In observation of children within this school, all of whom were enthusiastic computer users, she isolated two programming styles; hard and soft programmers. Although she observed that boys tended to differ from girls in their approach to computers (the boys were predictably described as the hard programmers) she concluded that, by using a flexible language such as LOGO, girls became as equally fascinated as boys.

Turkle described the hard programmers as those who needed control over the machines, whose aim was the completion of a premeditated, structured plan. These were seen as the scientists and the technologists, who in tests were shown to have an internal locus of control and a field-independent cognitive style. The soft programmers, usually female, showed the converse test results and were described as more artistic in their dealings with the computer. They experimented with their programs, and via their interaction with the computer allowed the program to develop in a manner 'more like a conversation than a monologue'.

However, Turkle's types bore no relationship to the differences between the sexes observed by the teachers of the Leicestershire schoolchildren, none of whom were programming in LOGO. The British boys demanded understanding and control over the machine *and* experimented with their programs, exploring the system with little extrinsic purpose. The girls were more concerned with the need for a final end-product from their efforts, but were frustrated by the methods which they had to employ when programming in BASIC.

LOGO has now crossed the Atlantic and is becoming very popular in some of our primary schools, as I witnessed when attending a lecture given by Papert to British teachers. A following of almost cult proportions had developed in his wake, and teachers of LOGO were unequivocally enthusiastic about its principles and applications. Unfortunately LOGO is rarely used in secondary schools, although young children growing up with this language may prove to have more positive attitudes towards new technology. Only time will tell whether the use of this language will remove the sex bias which seems so entrenched in the current system.

The schools' survey, initiated to try to discover why more females had not responded to the publicity, illustrated that females were not as intrinsically interested in machines and artefacts as males, merely in their usefulness. If machines such as computers prove difficult to use and are seemingly impractical, negative attitudes seem to be formed which reinforce the female's lack of ability in comparison with males.

To measure the validity of the teachers' theories and to gain some understanding of the use of home computers by children, about which the teachers knew little, a questionnaire was given to schoolchildren to investigate whether the differences in use and interest levels were substantiated by the boys and girls themselves.

The opinions of schoolchildren

Of the literature which discussed the disaffection of girls towards computers none was found which investigated the children's attitudes directly. No-one appeared to have asked children to specify details of their personal interactions with, and their interests in, computers and the second phase of this survey was intended to rectify this omission. The aim was to discover if the children's responses supported the conclusions that there were real differences between the sexes in their interaction and experiences with computers.

A multi-choice questionnaire was designed which could be completed by children from the ages of 11 years upwards. In order for as many children as possible to complete the questionnaire it was kept, on the advice of head teachers, very concise. This ensured that it would not create major disruption to timetables nor prove difficult to administer. The children were asked for basic demographic particulars, as well as details about their use of and interest in computers at home and at school, both in the classroom and in the computer clubs.

Children from four co-educational schools took part in the survey. These schools were selected to be as representative of the whole county as possible, and included:

School 1 – a rural upper school, with a very large catchment area, with children ranging from 14–18 years of age;

School 2 – a secondary school in a market town, with an age range from 11–18 years;

School 3 – a rural middle school, with an age range of 11–14 years;

School 4 – a city secondary school, with an age range of 11–18 years.

The teaching staff in the four schools were asked to administer the questionnaire, and their cooperation was sought through discussions about the research held during staff meetings. It was not felt necessary for all the children in each school to take part, but teachers were encouraged to get at least two complete classes from each year group to complete the questionnaires. These were usually administered during registration periods when whole classes were assembled.

The number of children on roll in the four schools totalled 2720 and completed, unspoilt questionnaires were received from 1656; a response rate of 61 per cent. Questionnaires were completed by 886 boys and 770 girls, with boys forming 53·5 per cent of the sample. The majority of the analyses carried out on the raw date concentrated only upon differences between the sexes, although differences between the schools were examined when large variation was observed.

The teachers believed that part of the differences observed between the sexes when interacting with computers was due to the fact that girls did not see the relevance of computing in their lives and were therefore less interested. The children's reported interest levels are shown in Table 5.1.

Table 5.1 Frequencies of reported interest levels in computers

Reported interest in Computers	Boys	Girls	Totals
Very interested	318	86	404
Quite interested	373	360	733
Not really interested	195	324	519
Totals	886	770	1656

Although the largest proportion of both sexes reported their interest level as 'quite interested', 36 per cent of the boys and only 11 per cent of the girls stated that they were 'very interested' in computers. Conversely,

22 per cent of boys but 42 per cent of girls described themselves as 'not really interested'. Analysis of the frequencies revealed a significant difference between the sexes in their reported interest level in computers ($p < 0 \cdot 001$), confirming the observations of the teachers.

Both sexes from each school listed the same school subjects in which they were able to use the computer in class, suggesting that there was equality of opportunity for both boys and girls to use computers if they so wished. Nevertheless data revealed that far fewer girls took advantage of this opportunity, and many tended not to take subjects within which computers were used. Thirty-three per cent of the boys but 53 per cent of the girls had not used computers within a classroom situation, a significant sex difference, with $p < 0 \cdot 001$.

When the data were analyzed by school rather than sex, it was apparent that there were significant differences between the schools in the proportions of the children who had used computers, an occurrence which had little to do with the proportion of computers to children within each school. School 3, with the lowest ratio of computers to children, was shown to have a significantly higher frequency of children using computers than the other schools, and School 4 (with the highest ratio of computers to children) had a significantly lower frequency of classroom use than the rest, with both results at the $p < 0 \cdot 001$ level (see Table 5.2).

Table 5.2 Children using computers in the classroom, and ratio of computers to children, by school

Computer Use	School 1	2	3	4	Total
Questionnaire Sample					
Number of respondents	550	384	186	536	1656
Had used computers in class	351	262	152	196	961
Percentage of sample	63·8	68·3	81·7	36·6	58·0
Whole School					
Total children on roll	1030	770	440	480	2720
Numbers of computers	5	17	2	6	30
Ratio computers:children	1:206	1:45	1:220	1:80	1:91

Even in their previous schools, (which depending upon the age of the child was either a junior or a middle school, neither of which prepare children for computer examinations) the sex difference in computer use was still apparent. Forty-one per cent of the boys but 34 per cent of the girls had used computers, a difference between the sexes significant with $p < 0 \cdot 003$.

The perceived link between mathematics, science and computing and 'masculine' interests was also investigated. The children were asked to list their 'favourite' and their 'least liked' school subjects, with both questions open-ended to allow total freedom of choice. The resulting data were analyzed to see how many of each sex listed computing or mathematics in response to either question (see Table 5.3). Science was unable to be isolated as in some schools science was combined with craft subjects. (N.B. Although the girls' results within each section are identical, these are correct and do not represent the same individuals.)

Table 5.3 Computing and/or mathematics cited as a favourite or a least liked school subject, by sex

School subjects	Boys	Girls	Totals
Favourite subjects			
Included computing/maths	395	257	652
Did not include computing/maths	491	513	1004
Totals	886	770	1656
Least liked subjects			
Included computing/maths	212	257	469
Did not include computing/maths	674	513	1187
Totals	886	770	1656

Over two-thirds of both sexes included mathematics and/or computing in answer to one of these questions, indicating that these subjects were on the whole either very popular or very unpopular, depending upon the child. Nearly 45 per cent of the boys and 33 per cent of the girls had spontaneously listed either or both computing and mathematics as favourite school subjects. This was a higher proportion of girls than had been anticipated from the teachers' reports, although once again the difference between the sexes was highly significant, with $p<0 \cdot 001$. The same significance level was observed between the frequencies of boys and girls who included computing or mathematics within their list of least liked school subjects, this time with 50 per cent of the girls and only 24 per cent of the boys listing them.

The teachers had made it very clear that girls rarely, if ever, made use of the computing facilities during the club sessions at lunchtimes and after school. In certain instances it was reported that first-year girls would make the attempt but very quickly ceased attending because of competition with the boys. The results confirmed that the boys were more likely to take the opportunity to use the clubs, with 33 per cent of the boys and only 12 per cent of the girls ever having attended (a difference significant at $p<0 \cdot 001$).

All the results from the questions pertaining to computer use within schools indicated that the boys were significantly more interested in and more likely to use computers than girls, confirming the reports from the computer studies teachers. The remaining questions dealt with computer use at home, an area about which the teachers had been unable to give any specific information.

Home computer ownership was very high. The results showed that 44 per cent of all the children questioned had a microcomputer at home, although the ownership was not evenly distributed by school or by sex, (see Table 5.4). The sex differences were still very apparent. Fifty-three per cent of the boys but only 32 per cent of the girls had computers at home, and the analysis showed this difference to be significant, with $p<0{\cdot}001$.

Table 5.4 Home computer ownership, by school and sex

Ownership	School 1	2	3	4	Total	Boys	Girls	Total
Owners	267	150	77	227	721	471	250	721
Non-owners	283	234	109	309	935	415	520	935
Totals	550	384	186	536	1656	886	770	1656

School 1 showed a significantly higher frequency of home computer ownership than the other three schools, ($p<0{\cdot}01$), but as this was an upper school only taking children from the ages of 14–18 years the age range was thought to account for this difference. Older children were perhaps more likely to be considered responsible owners by their parents, and were more likely to need the use of a home computer for examination work.

To see whether the possession of a home computer was correlated with the children's reported level of interest in computers, the interest levels were subdivided between those who owned computers and those who did not, and between the sexes within those two groups, (see Table 5.5).

Table 5.5 Interest levels in computers by ownership and sex

Interest level	Very	Quite	Not really	Totals
Computer owners	300	325	96	621
Non-owners	104	408	423	935
Owners—boys	260	181	30	471
Non-owners—boys	58	192	165	415
Owners—girls	40	144	66	250
Non-owners—girls	46	216	258	520

The owners of computers reported significantly higher levels of interest in computing than the non-owners ($p<0{\cdot}001$). One could assume that their interest had encouraged the purchase, or that their interest had grown with its use. The higher interest of the owners was also observed when the data from each sex group were analyzed, with the results from the boys significant at $p<0{\cdot}001$ and the girls at $p<0{\cdot}01$.

Of the children who reported themselves to be very interested in computers, 82 per cent of the boys had a computer at home, while only 47 per cent of the girls who stated that they were very interested had one, a significant difference of $p<0 \cdot 001$ between the sexes. In addition the male owners were more interested than the female owners ($p<0 \cdot 001$), and even the non-owning boys were more interested than the non-owning girls ($p<0 \cdot 01$).

All non-owning children were also asked if they would like to own a computer, and once again it was the boys who were more likely to want to own one than the girls, ($p<0 \cdot 05$). However, there was one result which showed no significant difference between the sexes. Of the non-owners who had no desire to own a computer, both boys and girls exhibited similar low levels of interest in computers (see Table 5.6).

Table 5.6 Non-owners of computers who did and did not want to own a computer, by sex and interest level

Non-owners of computers	Boys	Girls	Totals
Did want to own a computer	231	251	482
Did not want to own a computer	184	269	453
Interest of those not wanting computer			
Very interested in computers	2	1	3
Quite interested in computers	45	55	100
Not really interested in computers	137	213	350

To investigate just who had been influential in obtaining the home computer the children were asked to list the family members who had been 'keen to get the computer'. These lists could obviously include parents, siblings, as well as self where appropriate. These data were initially divided into three groups; those made up from male family members only, those which were only female, and those which included members of both sexes. The results showed that of the 721 children who had computers at home, 533 of the computers had been wanted only by male members, 142 by mixed groups, and only 46 by female members alone, confirming the sex bias.

The data were then analysed to see if the children had included themselves amongst those who had wanted to purchase the computer. The vast majority of the boys, 78 per cent, had included themselves within the list, but only 36 per cent of the girls had done so (see Table 5.7). Analysis of these data yielded a significant difference of $p<0 \cdot 0005$ between the sexes.

To examine how much spare time the children devoted to their computing, the owners were asked to estimate the number of hours they spent each week upon the home computer. The responses from the boys and girls are given in Table 5.8; divided into three sets for analysis.

Table 5.7 Children who did and did not want the computer, by sex

Who wanted computer	Boys	Girls	Total
Did include self	368	89	457
Did not include self	103	161	261
Totals	471	250	721

Table 5.8 Reported time spent on home computer per week, by sex

Time spent on home computer	Boys	Girls	Totals
No time spent	52	77	129
Some time spent	419	173	592
0–9 hours	330	234	564
>10 hours	141	16	157
0 hours	52	77	129
1–2	66	89	155
3–9	212	68	280
10–19	83	13	96
20–29	38	2	40
>30	20	1	21

Although one could assume that these figures included a great deal of inaccuracy, the teachers had agreed to ask the children to try to estimate their hours very carefully. The teachers' subsequent reports indicated that many children had pondered long and hard, and had asked for assistance in some instances. The results indicated that 11 per cent of the boys and 31 per cent of the girls did not use the home computer at all, and that 30 per cent of the boys but only 6 per cent of the girls used the computer for more than 10 hours per week. When the figures were broken down into small categories it was possible to see that the majority of children who used their home computers for many hours each week also tended to be male. The results from these analyses yielded significant differences between the sexes, with $p<0 \cdot 001$ in all three cases.

All of the results, whether to do with school or home computing, confirmed that there were indeed very significant differences between the boys and girls in their interest in, and use of, computers.

Discussion of the children's data

The children's data confirmed the opinions of the teachers and the literature that there were very significant differences between the sexes in their attitudes towards, and their activities on, computers. Girls reported themselves to be significantly less interested in computers than boys, as was born out by their responses to all of the questions.

A significantly greater number of girls than boys cited computing and/or mathematics as among their least favourite subjects; subject areas in which computers were more likely to be employed than any other. The fact that both sexes were offered computing in the same range of subjects, although fewer girls than boys ever used computers in the classroom, indicated that girls were deliberately opting out and choosing not to use computers. Although causality cannot be established, the implied links between computing and mathematics and the theory that these subjects are considered to be 'masculine' appear to be born out.

The data also revealed that not all of the children who had expressed an interest in computers had yet been able to use them in the classroom. One can therefore conclude that not only were girls opting out of computing more frequently than boys, but also that there was perhaps lack of opportunity for either sex to use them in some cases. School facilities and organization obviously play a great part in this situation.

School 3, a middle school, had a significantly higher proportion of children using computers during class-time than the other schools in spite of the fact that they had the lowest ratio of computers to children. There was no computing department as such, and their two computers were kept on trolleys allowing them to be wheeled to classrooms as required. The computers were only used in the context of other subjects, usually as a vehicle for commercial software, and no programming was taught. The teacher responsible for the computers, who believed that children should learn about them in context rather than as a special subject, had assured parents that as many children as possible would have the opportunity to use computers in one subject or another. He seemed to be achieving this end, as 82 per cent of children in the school had used computers. However, in spite of his ideals, the girls in his school were no more likely to be interested in computers than those from the other three schools which did not follow this philosophy. Mere exposure to computers, even if being used as tools and not for learning about the hardware and programming, did not appear to be sufficient to warrant a change of attitude in the girls. This attitude seems to have been formed early in life, as the results showed that the sex differences already existed during their junior school years.

Home computing, well removed from the competitive, masculine arena of the school computer room, did little to adjust the imbalance observed between the sexes. At home, although children were free from academic constraints and able to explore the computer's capabilities in a non-threatening environment, girls still showed similar reluctance to use computers. Fewer girls had a computer at home, and those who had were less likely to have desired its purchase and less likely to be interested enough to use it for more than brief periods of time. Although the majority of boys who stated that they were very interested in computers had a computer at home, this did not apply in the case of girls. Their interest did not necessarily lead to ownership, indicating that the applications for which one could use a home microcomputer were perhaps not considered useful to them. From the results one could assume that rather than doing something to the computer the girls wanted to make it do something useful for them.

Although the schoolchildren were not asked questions about computer dependency, the results from their reported estimates of time spent on the home computer each week gave indications that dependency might have been apparent in some children. Twenty-one children estimated that they spent over 30 hours each week computing, a large proportion of their spare time. Only one of the children reporting this level of computer use was female, once again indicating that intensive use of computers was a male-dominated activity.

One may conclude from the data that there were very real and significant differences between the sexes in their use of computers, even in an environment where both boys and girls have grown up with new technology and were free to use them with equal frequency and for similar tasks if they so desired. Even when great efforts were made to ensure that computers were used as tools and where computing was not taught as a subject in its own right, the sex differences remained. Without a basic interest in computers one is unlikely to wish to learn more, and without this desire computer dependency can hardly follow.

It was not my intention during the time available for this survey to offer solutions to the problems encountered within schools, (for discussion of these areas see Saunders, 1975; Kelly, 1982; Mahoney, 1985; and Ward, 1985) but to determine whether the sex differences observed in the ratio of men to women volunteering to take part in the research were apparent within the controlled educational environment. If in this situation no differences had been observed one could conclude that mere lack of opportunity and differentiation of computer tasks between the sexes were the reason why so few women responded. However, the results from the school children indicated that this was not so.

Discussion

The responses to the publicity from the general public yielded a sample of people of whom few were female; too few in fact to allow their subsequent data to be analyzed separately. When females are under-represented within a group under study their data is frequently ignored as unimportant (see Borrill and Reid, 1986). However, the fact that they were so heavily underrepresented was a very significant result, and merited its own investigation.

Two major factors were thought to be responsible for this discrepancy between the sexes. Women were considered to have had less opportunity than men to use computers while at work, and if used they were thought to use them in a less creative manner and with less autonomy than males. Both could be seen to inhibit women from viewing the computer as a machine for leisure and enjoyment, and as a consequence they would have been less likely to have become dependent upon them.

A secondary school population was specifically selected to examine sex differences, because both boys and girls are offered equal opportunity to use computers for the same tasks while at school, thereby eliminating the factors considered to be instrumental in the adult population. Although seemingly non-discriminatory, the use of computers within schools and the interest levels observed still persisted in showing extreme differences between the sexes even within this young age group who had grown up with the technology. One can therefore assume that factors unrelated to opportunity were in effect, which could also have been active within the adult population. These differences appeared primarily to be due to a basic contrast of needs and interests, together with strong socialized sex-role stereotyping rather than to inherent differences in ability.

The results indicated that females were more likely to require a useful end-product from their computing efforts and expected computers to be more controllable and more easily used than in fact they were. Together with initially being more inhibited by the technology, they were more likely to doubt their capabilities to use the computer and were more easily frustrated when programs did not run at the first attempt. They expressed less desire to understand the basic workings of a computer, tended to see it primarily as a tool and expected it to produce the desired results rapidly without too much effort on their part. At the time of this survey it is fair to say that without expensive software, a printer, and adequate support material in the form of well-written manuals, there was little which could be easily accomplished on a microcomputer; therefore their needs were difficult to satisfy. There is no reason to believe that these theories apply any the less to adult females than to schoolgirls.

Informal discussions with women, most of whom regularly and efficiently use such modern technological devices as cars, washing machines, televisions and stereos and are not intimidated by them, often reveal that they have little desire to understand the technology but are more concerned with their applications. This of course applies equally well to some men, but there appears to be a greater likelihood that more males than females are also interested in the mechanical and electronic aspects of these artefacts.

Not wishing to conspire with such comments as 'It is not technology that is out of control, but capitalism and men' (Cockburn, 1985) when describing the male domination of new technology, there appears little doubt that the introduction of computers has largely been due to the efforts of men, and therefore tends to reflect male interests. In general, computer advertising is directed at boys and men, software whether for leisure or business is biased to male interests, computing magazines are designed for a male readership, and computer retail outlets are staffed by men; theories well-substantiated by Haddon (1986). Such factors were believed to alienate many females from this male domain even before they had any experiences of using computers, and, as with science and mathematics, feelings of helplessness and lack of confidence develop and inhibit the growth of interest.

Some people have begun to recognize that Britain is in danger of losing half its talent if females do not acquire computer skills, and companies such as Acorn Computers have initiated new publicity methods (cited by the Equal Opportunities Commission, 1985) in an attempt to address this imbalance. In addition, some software houses are now designing programs specifically for women, and some retailers are beginning to employ female assistants with the aim to encourage women purchasers, (Begole, 1984). It is not known how successful these ventures have been.

Even if interest is not initially destroyed in females, the uses to which microcomputers can easily be put tend not to be seen as relevant to their lives. Tittscher (1984) reported that 96 per cent of home computers were purchased by men and that 88 per cent of the users were estimated to be male. He believed that the low use by women was caused by the lack of practical applications to which computers could be put; an attitude which mirrored those of the schoolgirls as reported by their teachers.

Not all women conspire with these social pressures and a few do enter the world of computing, but it appears that many have found great difficulties when competing in this male-dominated sphere. *Microsyster*, a group of female consultants based in London, operates a help-line for women trying to enter the world of new technology, offering advice, encouragement and career guidance. Much of their work concentrates upon boosting the confidence of women, whose abilities are not in doubt, to enable them to compete on equal terms with men (Craig, 1986).

Lack of opportunity tends to prevail for women working in the computer industry; however, in certain circumstances women fair very well in the field of new technology, especially when their skills and needs are specifically catered for. *F International*, a British company set up by Ms. Steve Shirley in 1962, employs a unique workforce which relies upon homeworkers as programmers and data processors. The vast majority of the employees are women, working 20–25 hours per week from home via terminals. A personal interview with one of the area support managers revealed that women were preferred within the company as they had proved very reliable, maintained a high standard of output and fulfilled their time targets. Conversations with other employers and managers, employing both male and female programmers, frequently stated that women were considered to be more desirable because of their reliability and application to the task, while some men were criticized because of the unnecessary time spent performing tasks extraneous to those demanded.

These comments are reminiscent of the teachers' descriptions of the different methods of computing observed in boys and girls; much of the boys' time was spent refining programs without adding anything to the intrinsic quality of the output. If commercial programs are written by people with such interests and if the resultant software is difficult to use it could lead to the alienation of many potential users, especially women. The argument becomes circular.

This research has not shown that females are intrinsically unable to use, enjoy and benefit from new technology, but it does go some way to explain why women and girls were more likely to be reluctant to work with computers and were less interested in their use. When females did use computers, their modes of working appeared to differ from those of many of their male counterparts and their need for practical applications appeared to inhibit them from wishing to own a machine of their own. The schools' data revealed that far fewer females owned or wished to own a computer even if they had expressed an interest in them.

As the results from the Introductory Questionnaire indicated that ownership of a home computer appeared to be an almost essential prerequisite for the formation of computer dependency, this alone could explain the discrepancy between the sexes. Without a basic interest in new technology a computer would neither be purchased nor used voluntarily, and without this interest computer dependency would not follow. The results showed that the proportion of schoolgirls to boys becoming dependent upon computers (1 : 26) was not significantly different from that found in the responses to the publicity from the general public (6 : 100), and one may conclude that the sex differences in responses to the publicity were representative of the computer dependent population.

However, there was no reason to believe that the few females who were interested in new technology, who did own computers and did use them frequently, were any less likely than men to become dependent on them. From an analysis which compared the data from Tittscher's (1984) review of retail surveys and the data from this research, no significant difference ($p > 0 \cdot 05$) was found between the percentages of females to males buying computers, or being the main users, when compared with the frequencies of females to males reported to be dependent upon computers, at school or in response to the publicity (see Table 5.9). One may therefore conclude that of the few women who are intrinsically interested in computers, who do buy and use them, equivalent proportions of both sexes do become dependent upon them. The primary difference between the sexes centres on their willingness and need to use and buy computers in the first instance; a factor which seems to be in part culturally induced.

It appeared that the characteristics of the majority of computing activities were more appropriate and more likely to match the needs of

Table 5.9 Purchase, use and dependency upon computers, by sex

Aspects of computer use	Males	Females
Buyers of home computers (Tittscher, 1984)	96	4
Main users of home computers (Tittscher, 1984)	88	12
Dependent respondents from national publicity	100	6
Dependent children reported by teachers	26	1

the male than female users. It was therefore understandable that females were less likely to become fascinated with the intrinsic qualities of the computers and were less likely to use and buy them. When computers are used as tools, for example to analyze statistics or as word-processors, they are unlikely to produce the type of behaviour which could lead to dependency, as the computer is merely a vehicle for software and is in itself only subliminally acknowledged to exist. Conversely, when the computer is used in order to discover its capabilities and explore its potential, the computer becomes the focus of that activity and the principal object of study. Because of the wealth of activities which can be performed when writing one's own software, the computer can hold limitless potential for the people who exhibit this type of interest, who in the main seem to be male.

The survey into the children's attitudes to computers confirmed that the responses to the publicity did reflect the extent of the differences of computer dependency between the sexes. As a significantly greater number of males owned, used and were interested in computers, for a variety of well-substantiated reasons, one can conclude that maleness is one of the most significant variables in the development of the syndrome of computer dependency, simply because of the male's greater interest in new technology.

Computer dependency was observed to occur to only very few of all young computer owners, estimated to be one in 300 schoolchildren who owned a home computer; the following chapter investigates the nature of this phenomenon.

Chapter 6
Computing activities: The nature of the phenomenon

Introduction

This chapter concentrates upon the nature of the syndrome of computer dependency, and deals with the investigations carried out to discover the computing activities undertaken by the people who considered themselves to be dependent upon computers, the Dependents. Comparisons are made between their responses and those of the control group of non-dependent computer owners, the Owners. The aim was to determine if the computing activities of the Dependents were in any way unique and different from those of the Owners. Although it was anticipated that there would be differences in the time spent and frequency in the use of the computer it was not known whether their activities would also differ qualitatively.

The interviews with the undergraduate students had indicated that those who described themselves as dependent often spent their time exploring the system rather than aiming to achieve a practical end-product from their programming. However, the results from the schools' survey suggested that this type of computing was perhaps merely typical of male computer users in general who exhibited different attitudes and interests from females.

By comparing the computing activities of the Dependents with the Owners, the majority of whom were male, it was therefore anticipated that, because of the matching process between groups, fewer differences would be found than between the male and female school pupils. However, if significant differences were found between the Dependents and the Owners the variables isolated could be considered relevant to the syndrome of computer dependency and not merely to differences between the sexes.

Methods employed

The data within this section were obtained from questions included in the General Questionnaire, the Computing Questionnaire and the Final Questionnaire. Comparisons were made between the responses from 75 of each of the two groups of computer owners, the Dependent and the Owners, and further clarification was obtained from the interviews with 45 of the Dependent group.

The chapter is divided into sections which address the following questions:

(1) Were there differences between the two groups of computer owners in their initial introduction to computing and programming, and could these somehow account for the Dependents' later dependency?

(2) Did the time devoted to computing differ quantitatively and qualitatively between the Dependents and the Owners?

(3) Were there quantitative differences in the computer hardware owned by the two groups which could lead to qualitative differences in the activities undertaken?

(4) Were some computing activities more likely to be undertaken by one group than the other and did their interest in, and attitudes towards, their computing activities differ qualitatively?

Each section contains a summary table of the questions asked from which the comparisons between the Dependents and the Owners were made. The significance levels of the results are annotated by asterisks as follows:— *** = $p<0{\cdot}001$, ** = $p<0{\cdot}01$, * = $p<0{\cdot}05$, – = $p>0{\cdot}05$, and the text describes the direction of these differences. The small number of women within each group made formal comparisons between the sexes inappropriate.

Initial experiences with computers and programming

Were there differences between the groups of computer owners in their initial introduction to computing and programming, and could this somehow account for the Dependents' later computer dependency? Table 6.1 summarizes the questions which addressed this area.

The majority of the individuals within both groups of computer owners held the preconceived idea that they would make either good or fair programmers even before they had had the opportunity to use a computer. Only 10 per cent of either group thought they would be poor, and there was no significant difference between the results from the Dependents and Owners. Neither was there any significant difference between the

Table 6.1 Initial computing and programming experiences

Initial computing and programming experiences	Significance level
Preconceived ideas of ability to use computers	–
Year first used a computer	–
Year first bought a computer	***
Computers used in education or at work	–
When learnt to program	–
Was programming self-taught or instructed	***

two groups for the reported year in which they had first used a computer, with approximately half using them before the first cheap microcomputers became available in 1980. Ownership of computers naturally tended to come later than first experiences for many people, but in this instance the Dependents were significantly more likely to have bought their personal computer earlier than the Owners. Although the majority of both groups did not purchase a machine until the microcomputer boom, almost twice as many Dependents as Non-dependents had purchased one before 1980 when microcomputers were scarce and very expensive.

Although the majority of both groups had used computers of some description either within a working or an academic environment, the majority of the Dependent group (83 per cent) considered their programming to have been entirely self-taught, while most of the Owners (58 per cent) had been totally or partially taught to program by others. When asked how quickly they felt themselves to have become dependent upon computers, 71 per cent of the Dependent respondents stated that this had occurred immediately upon use of a computer, with the remainder reporting that this had taken a few days or weeks. None claimed it took months or years.

Discussion

Both groups of respondents appeared to have had equal opportunities to use computers in working or academic environments, no doubt a reflection of their technological and scientific backgrounds, and most considered that they would show an aptitude for computer programming. Such results appeared to confirm the teachers' theory which concluded that the computer was not a machine which caused apprehension to the majority of males. However, a significantly larger proportion of the Dependents had purchased a personal computer before the Owners, had taught themselves to program and had almost immediately become dependent upon them.

Further clarification of the Dependents' early interest in computers was revealed during the interviews, with quite a number stating that they had been 'hooked' even before they used a computer and were merely waiting for the opportunity to have direct experience. It appeared that many were not only already knowledgeable about computer technology but also that some had studied the structure of programming languages even before they had had the opportunity to use a keyboard. Their first experiences often merely confirmed their previous fascination and gave it a more concrete basis.

> 'The first day I had the *TRS 80* I sat down at 6 pm and the next time I looked up it was 3 o'clock in the morning.'

It was not unusual for new owners to spend the whole of the first few days experimenting with their computers to find out what they could do, often forfeiting meals and sleep in the process. The Dependents were not deterred by 'unfriendly' systems and poorly designed manuals, as are so many other naïve users, but appeared to delight in finding their way about the computer system through their own efforts, learning from their mistakes as they progressed. They already had an awareness of the power and potential of computers, often seemingly enhanced by their readings in science fiction as well as from factual material, and often stated that their interests in new technology were simply a natural extension from their other hobbies and interests, which for many tended to be electronically- and technologically-based.

With interest aroused long before they had the opportunity to use a computer, many Dependents purchased their own machine as soon as they were able to do so, in order specifically to learn about computers and programming methods. It was therefore understandable that programming was invariably self-taught for this group. The interviewees described their first experiences with the computer as exploratory in nature; their aim was to discover the capabilities of the computer and to use it as a vehicle for self-education rather than to undertake a specific task or problem-solving exercise.

In contrast with the Dependents' initial experiences, one may deduce from the results that the Owners were more likely to have had their interest aroused from using computers in a functional setting. Only after this experience, which tended to include specific training in programming, were they likely to purchase a computer. Their introduction to computing would therefore not need to be exploratory, and being based upon previous experience was more likely to be task-oriented.

The differences in users' introduction to computers and programming may therefore be seen significantly to affect their attitudes to new technology and its subsequent use. The results seemed to indicate that one was more likely to become dependent upon computers if the first experiences were self-motivated, based upon previous interest and from the desire to learn about computers first hand by exploratory, trial-and-error techniques. A more structured introduction may lead to the computer being viewed as a tool rather than primarily as a vehicle for a new learning experience, and such an introduction can be seen to be more likely to inhibit the development of computer dependency.

Characteristics of time spent computing

Did the time devoted to computing differ quantitatively and qualitatively between the Dependents and the Owners? Table 6.2 summarizes the questions asked to investigate this area.

Table 6.2 General aspects of time spent computing

Time spent computing	Significance level
Hours spent computing per week at home	***
Hours spent computing per week at work or study	***
Hours spent computing per week both at home and at work	***
Maximum time spent at any one sitting	***
Days per week that home computer is used	***
Percentage of spare time spent on home computer	***
Find it difficult to stop when computing	***
Lose track of time when computing	***
Attempt to ration time spent computing	–
Have been rationed by others	–

As anticipated the Dependent group did spend significantly more hours on their computers and used them more frequently than did the Owners. The estimates made of the time spent computing per week, whether at home, at work and study or in total, were significantly greater for the Dependent computer users than the Owners, as were the estimates for the longest time spent computing at any one sitting without a proper break for meals or rest. The means and standard deviations of the estimated times from both groups are given in Table 6.3.

Table 6.3 Estimated hours spent computing. N, number of subjects; S.D., standard deviation

Time spent computing per week in hours	Dependents			Owners		
	N	Mean	S.D.	N	Mean	S.D.
At home	74	23·6	12·4	75	6·0	4·4
At work/study	53	20·7	16·6	57	10·7	10·6
Total hours	75	37·9	18·9	72	13·8	11·0
Maximum at 1 sitting	75	11·8	9·9	73	4·4	3·6

There was extreme variation in the responses within each group when asked to estimate the times they spent computing, with the maximum total time given at 80 hours from one Dependent who computed both at home and at work. At home, where one has more autonomy and control over the hours spent, the mean estimates from the Dependents were almost four times those of the Owners; at work the Dependents spent twice as long; and in total almost three times as long as the Owners. Even allowing for a certain amount of inaccuracy these differences are highly significant.

When asked to state the maximum time they had spent computing without a proper break for food or rest, the mean estimates from the Dependents were almost three times greater than those of the Owners, with the longest reported time spent being 48 hours. One Dependent respondent, who spent over 45 hours of his spare time per week on his home computer, unexpectedly gave almost the shortest maximum time at one sitting, just three hours. However during the interview he explained that he took regular breaks because he thought overheating the computer could be detrimental to its functions. He stopped for the computer's sake, not his own.

From these results it was understandable that the Dependent group also reported using their home computers with greater frequency than did the Owners. Estimating the numbers of days per week they used their home computers, if only for a short period of time, the majority of the Dependents used them daily. Their results yielded a median of seven days per week and the Owners only one.

When asked to report the percentage of their spare time which was devoted to computing, the Dependents and their spouses both estimated this to be well over 50 per cent, while the Owners only considered that they spent an average of one seventh of their spare time at the computer, a difference which also proved highly significant (see Table 6.4).

Table 6.4 Estimates of the percentage of spare time spent computing.

Percentage of spare time computing	Dependents			Owners		
	N	Mean %	S.D.	N	Mean %	S.D.
Respondents' estimates	75	56·2	23·2	71	15·3	14·7
Spouses' estimates	22	63·9	13·9			

Although it was apparent that the Dependents reported spending more hours and a greater percentage of their spare time computing than the Owners, this could have been due to the fact that for some reason they had considerably more spare time than the Owners to spend upon this activity. A simple computation was carried out to estimate whether there were gross differences in the spare time that each group claimed to have. By converting the data from each group for the mean hours per week spent computing at home, and the mean percentages of spare time this involved, it was possible to determine the total spare time that each group considered they had at their disposal (Table 6.5).

Table 6.5 Calculated total spare time available per week

	Mean time on home computer	Percentage of spare time on home computer	Mean spare time per week
Dependents	23·6 hours	56·2%	41·9 hours
Owners	6·0 hours	15·3%	39·2 hours

The mean results showed that both groups estimated they had approximately 40 hours of spare time per week. Although the Dependents' estimate was slightly greater than the Owners', this small difference could not account for the gross discrepancies between the groups. This result confirmed that the Dependents chose to devote a larger proportion of their spare time to computing than to other activities, as one would expect of people who were exhibiting symptoms of computer dependency.

The interviewed students had stated that they frequently found it difficult to stop computing once they had started and that they tended to lose all track of the passage of time; for this reason some had attempted to ration themselves or had been rationed by others, often by the university's computer centre. The results to questions investigating these areas indicated that the Dependent group likewise were significantly more likely than the Owners to find it difficult to stop computing once they had started and that they frequently lost all track of time when at the keyboard. However, few of either group had attempted to ration their time spent computing or had been rationed by others, and no significant differences were observed. The fact that most of the respondents, in both groups, were adult may perhaps explain this. Although many of the Dependent interviewees revealed that their spouses had made some attempts to curb their computing, their efforts had largely been unsuccessful.

Discussion

The results yielded very significant differences between the Dependents and the Owners for the estimated times spent computing, in all instances showing the Dependents devoting more of their time to the activity. When spouses of the Dependents were interviewed the impression was gained that some of the Dependents had under- rather than over-estimated the times they spent, stating that they rarely saw their husbands after they had returned from work and had eaten dinner, when they retired to their computers there to remain until the early hours of the morning. That extremely long hours were spent at one sitting was demonstrated when visiting one of the Dependents with whom an interview had been scheduled for nine o'clock in the morning. Before we could commence the interviewee excused himself to visit the local shop, bringing back two packets of peanuts which were to constitute his breakfast: he had been programming all night with no breaks of any kind.

The majority of the interviewees indicated that they would spend far more time computing if they did not have to sleep or go to work, with many stating that there were not enough hours in the day to undertake all the computing that they wished. They did not therefore consider the time they spent computing to be excessive. Not only at home were spare hours spent computing; those who had access to computers at work frequently stated that they forfeited their lunch hours and breaks to undertake their own computing, with one stating that he 'played' with the computer when he should have been working. From these reports it came as little surprise to discover that some people had even taken holidays from work in order to do more computing, with one man taking a week off to reach 'Wizard' status on an adventure game. If regular holidays were customarily taken away from home some took their computers with them, although for others this proved impractical:

> 'When I went on holiday I couldn't stop thinking about it—it felt so great when I got back. They were real withdrawal symptoms. I'll have to take an enormous pile of computing books and magazines with me if I'm ever going to cope without my computer again'.

When asked to state what they would be doing if not computing, many Dependents thought they would return to doing odd jobs about the house and garden or would spend more time with their families, which they admitted now tended to be sorely neglected. Others thought they would devote more time to reading, watching television, or undertaking their previous hobbies which were now largely ignored, but were adamant that none of these activities could offer the fulfilment found when computing.

One may conclude that the time the Dependents spent at their computers was atypical of other computer owners. They were not only prepared to spend extremely long hours computing, but some would forego sleep, food and holidays for the ability to increase this time. Even past hobbies, which had previously been of immense interest, were now abandoned for computing which they considered to be far more intriguing. The results from the questionnaires and the interviews revealed that for all but one of the Dependents the computing which brought them their greatest fulfilment was carried out at home on their personal computers (only one did not own a computer but programmed solely at work). At home they could undertake their preferred activities without restriction, and as such these activities were more representative of their dependency. For this reason much of the remainder of this book tends to concentrate solely upon the computing activities carried out at home.

The hardware involved

Were there quantitative differences in the computer hardware owned by the Dependents and the Owners which could lead to qualitative differences in the activities undertaken? Table 6.6 summarizes the questions from which comparisons were made between the groups.

Table 6.6 Ownership and understanding of computer hardware

Computer hardware	Significance level
Number of personal computers owned	**
Modems owned, to access networks	***
Need to understand the hardware	**
Value of computer hardware	***

The Dependents were found to own significantly more computers than the Owners, with the mean number of machines per group at 1·7 and 1·2 respectively. While only 21 per cent of the Owners possessed more than one computer, 37 per cent of the Dependents were multi-owners and 17 per cent had three or more machines. Significantly more of the Dependents (77 per cent) than the Onwers (38 per cent) also owned modems which gave them access to other networks and databases.

When asked whether they had a need to understand their hardware or not, the results revealed that significantly more of the Dependents (64 per cent) reported a need for technological understanding of their equipment than did the Owners (43 per cent). The Dependents' greater interest in their hardware was also reflected by the amount of money they invested in computers and peripheral equipment. Even though the Owners' estimates of the value of their hardware were high, at nearly £1,000, the Dependents' estimates were almost double this (see Table 6.7). There was great variation within each group but the differences still proved highly significant.

Table 6.7 Reported values of computing hardware

Group	Range of values	Mean hardware value	S.D.
Dependents	£150–15000	£1832·7	2210·9
Owners	£ 30–6000	£ 996·3	1186·7

More details of the Dependents' hardware were gained from the reports from the 106 respondents who completed the General Questionnaire, which asked for the make of their first computer and the equipment that they presently owned. The initial purchase for 38 per cent of them was either the *Sinclair ZX80* or *ZX81s*, the first cheap microcomputers available in Britain, while 27 per cent had bought the more powerful *Sinclair Spectrum* and *Acorn BBC* computers. Five had invested in the more expensive *Apple II* and two people had built their own systems. The remaining third of the group were divided between 19 various other models, some of which had been available in the late 1970s. These varied from expensive, powerful microcomputers to small games machines.

The Dependent interviewees revealed that the choice of their first computer depended simply upon its availability and price. However, many were disappointed by the lack of power of their first machines, and as an indication of the seriousness with which they took computing had quickly purchased other models. The results showed that 76 of the 106 Dependent respondents had bought other computers since their first purchase. Most had graduated from the less powerful *ZX80s* and *81s* to machines with greater memory, and by 1984 88 per cent owned either a *Sinclair Spectrum* or an *Acorn BBC B*. A few had invested more heavily, and other computers listed included *Compaqs, Tangerines, Apricots* and a *Wren Executive System*; all expensive systems.

Some seemed to collect computers as others collect cars. One single man owned an *Acorn Atom*, a *Nascom 1* ('an open-frame computer so that I can get at the components') and a *Micotan 65*, as well as two disc drives, two monitors and two modems. A married man in his forties who had two houses, one in Britain and one abroad, owned four *Apple* computers, two for each house so that a backup was always available in case of breakdown. Another owned a *Tandy*, a *Sinclair Spectrum*, a *Sinclair QL*, an *ITT Extra* and a portable *Tandy 100*, to ensure that he was never far away from a machine.

Few of the Dependents appeared to be content to use the computer as it stood and perusal of their listed hardware revealed that the majority had added many peripheral items of equipment. These included second processors to increase the capacity of the computer, disc drives to facilitate faster processing, modems to enable them to communicate with other networks, and printers so that they could obtain 'hard copy' of their programs, together with monitors, joysticks, etc. Many of these individual hardware peripherals cost more than the basic computer and it was therefore of little surprise that the estimated values of their hardware were large. As suggested by comparisons made with the Owners, most of the Dependents were technicians as well as users; they needed to understand the workings of their hardware and the adaptation and enhancement of their computers expressed this interest.

Discussion

These results confirmed the intense levels of interest shown by the Dependents in computing. They had purchased their first computers as soon as they were readily available and significantly earlier than had the Owners. That their interest had been sustained was demonstrated by the quantity and value of their present equipment. The interviewed Dependents, even if married, appeared to have little problem justifying the amount of money spent on their hardware and seemed to believe it was little different from that which others spent on their hobbies. 'It's just part of my "toy" budget—others have sports cars' seemed to typify many of the responses. Some had in fact made sacrifices to buy their equipment and one young man (who owned the most expensive system) stated that he used to save £200 per month and live on £50 in order to buy his equipment.

More of the Dependents than the Owners were in possession of facilities, such as modems and printers, which enabled them to attempt and undertake a wider range of computing activities. However, while most of the Dependents had invested heavily in their equipment and showed a need to understand their hardware, some owned only modest equipment which they used as it stood, showing similarity with many of the Owners. Such results suggested that computer dependency could be expressed in a variety of ways and was not necessarily confined to one particular type of activity.

Types of computing activities undertaken

Were some computing activities more likely to be undertaken by one group than the other and did their interests in, and attitudes towards, their computing activities differ qualitatively? Questions examining this area were confined to activities carried out on home computers only, where full autonomy and choice were present. Because of the quantity of data in this section, they have been sub-divided and are summarized in Tables 6.8 to 6.10.

The results from a multiple choice question included in the General Questionnaire gave the first indication that the Dependents' computing activities differed from those of the Owners. When asked to specify upon which of four main activities they preferred to spend their time, significantly more of the Dependents selected exploration of the computer system and use of the networks than did the Owners, whereas they both used commercial software and wrote specific programs with equivalent frequency (see Table 6.8).

The differences between the groups were clarified by further questions included within the Computing Questionnaire and the Final Questionnaire (see Table 6.9).

The Dependents were shown to devote significantly more of their computing time to programming than did the Owners, most of which was exploratory and of a self-educational nature rather than for a specific

Table 6.8 Computing activities upon which most time was spent

Computing activities upon which most time was spent	Significance level
Exploration of the computer system	***
Use of computer networks	**
Use of commercial software	—
Writing specific programs for an end-product	—

Table 6.9 Programming activities and attitudes towards programming

Programming activities and attitudes	Significance level
Percentage of time spent programming	**
Percentage of programming time spent in exploration	***
Number of programming languages able to be used	***
Use of machine code and assembler	***
Preferred programming language (Basic or other)	**
Programming style—hands on or pre-planned	*
Enjoy debugging programs	**
Programs considered to be well written	**
Programs documented with reminder statements	—
Graphics and sound added for aesthetic reasons	—
Programming has useful end-products	—
Feel creative when programming	—
Mind 'reinventing the wheel'	—

end-product. The Dependents' estimates showed an average of 49 per cent of their time devoted to programming, almost two-thirds of which was exploratory, while the Owners spent 36 per cent of their time programming, one third of which was exploratory. Both of these differences proved to be significant. However, the responses within the Dependent group varied greatly with some devoting all of their programming time to exploration while others spent none. The two activities were found to be poorly correlated.

To indicate the range of their programming interests, the Dependents were shown to use more computing languages than the Owners. Only 17 of the Dependents used merely one language, and 19 were familiar with five or more. These frequencies were significantly more than for the Owners, three of whom knew no programming languages and 32 of whom knew only the high-level language BASIC, a version of which is supplied with almost all microcomputers. The mean number of languages used by each group was 3·2 for the Dependents and 2·0 for the Owners. Fifty-two of the Dependents, but only 28 of the Owners, were also able to use the low-level languages of machine code and assembler. These were the languages most frequently selected as those preferred by the Dependents, while the Owners who were familiar with two or more languages still preferred to use BASIC.

In addition to differing in the time they spent programming and in the languages used, the two groups of computer owners also differed in their adopted programming styles and methods. The majority of the Dependent group programmed 'hands-on', allowing the program to develop at will, with only 10 per cent stating that they programmed with a pre-written structured plan. The majority of the Owners on the other hand only programmed after they had carried out some preliminary planning. Without adequate planning programs often require considerable debugging but significantly more of the Dependents admitted positively to enjoying this aspect of computing than did the Owners. Although pre-planning was lacking from the Dependents' programming methods, this group was significantly more likely than the Owners to consider their programs to be well-written, perhaps because of the time dedicated to debugging and programming in general.

In spite of these differences in programming styles, both groups documented their programs with Rem (reminder) statements with equal frequency, with slightly more undertaking this activity than not. There were also no differences between the groups for those who added graphics or sound to their programs in order to improve the aesthetics, with the majority of both groups always or usually adding such embellishments. Whether dependent or not, the majority of both groups also freely admitted that most of their efforts yielded no extrinsically purposeful or useful end-product, but did consider that they were creative when computing. This did not seem to conflict with the fact that the majority of both groups did not object to 'reinventing the wheel' when programming.

Activities other than programming were also carried out by the groups, and these illustrated further differences of interest between the Dependents and the Owners (Table 6.10).

Table 6.10 Other computing activities undertaken

Other computing activities undertaken	Significance level
Time spent playing games	★★
Types of computer games played	–
Quantity of software owned	★★★
Software bought or copied	★★★
Adaptation of commercial software	★★★
Hacking into other computer systems	★★★
Attitudes towards hacking	–
Hacking (modem owners only)	–
Networking activities undertaken (modem owners only)	–

In addition to their programming activities, significantly more of the Dependents played computer games than the Owners. However, this was not especially widespread even among the Dependents, as only one third reported spending much time playing games. Nevertheless, the players within both groups enjoyed the same types of games, with adventure and arcade games being the most popular, followed by simulations of aircraft flightdecks and other vehicles. Few of either group listed the computerized versions of the traditional board games as their favourites; those of *Scrabble*, chess and backgammon, for example.

That the Dependents spent more time playing games could have been as a direct result of their collections of commercial software which were found to be significantly larger than those of the Owners. The Dependents were also significantly more likely to have copied the commercial software in their possession than were the Owners, in spite of the fact that software is subject to copyright and often protected by security devices. Not only were the Dependents more likely to 'pirate' their software but they were also more likely to adapt and alter these programs having broken the security codes. Similarly, the Dependents had significantly greater experience in 'hacking', illegally breaking into the computer systems and databases of others. However, when the results from the modem-owners were specifically isolated, equivalent proportions of both groups were shown to have attempted hacking, and the majority of both groups of computer owners considered hacking to be 'fair game', despite the illegality of the action.

Although networking was more likely to be undertaken by the Dependents because of their greater ownership of modems, of those who had this facility both groups enjoyed the same type of activities. The most popular use of the networks was for personal communication with other people using private 'mailboxes', closely followed by accessing general and computer information. Open communication with groups of people and accessing computer software were the least popular activities for both groups of modem owners.

The results suggested that some of the activities for which differences were found between the Dependents and the Owners were either software- or machine-driven. Without a modem one cannot undertake either networking or hacking, and without owning software, whether for business purposes or for games-playing, one cannot indulge in such activities. However, of the results obtained the majority did indicate that there were differences between the two groups of computer owners, not only quantitatively but also qualitatively, in the ways in which they used their home computers.

FOOTNOTE The term 'hacking' not only describes the process of breaking into computer systems, but is also used by others to describe the process of breaking software security codes and many American authors use the term to describe the general activity of exploratory programming, 'hacker programming'. However, to avoid confusion, the term 'hacking' is used in this book specifically to describe the activity of breaking into computer systems (unless quoting another author) as this tends to be the most accepted use of the term in Great Britain.

Discussion

As expected from the results from the schools' survey, which described the computing preferences of schoolboys, in quite a number of instances there were no significant differences between the Dependents and the Owners, all of whom owned computers and most of whom were male. The teachers' description of a positive 'masculine' orientation to computing were confirmed by some of the results in this section.

Both the Dependents and the Owners tended not to use their home computers solely for the purpose of writing specific programs, but were quite prepared to 'reinvent the wheel' by copying or rewriting the programs of others, an activity which encourages understanding of the processes of programming. Some played games, even the 'zapping spaceships' type of arcade games so popular with young boys. Both groups of modem owners used the networks for similar functions and were equally likely to have attempted 'hacking' into systems for which they had no legal right of access. From such results it was possible to see that the home computer could be viewed as an instrument for self-development and education, as a means of learning about new technology, but further results indicated that this view was more commonly held by the Dependents. The Owners seemed more likely to desire useful end-products, even if they were unattainable.

It was possible to deduce from the results that the Owners were more likely than the Dependents to have purchased their home computers with the intention of using them for practical rather than self-educational purposes. They were more likely to have had prior experience in programming before buying their own computers and were therefore already familiar with some of the real capabilities and limitations of new technology. They had received formal instruction in programming and even when working at home their programming style showed that they tended to pre-plan their programs rather than work 'hands-on', indicating that it was likely they had an end-product for their work firmly in mind. This, together with the results which showed them to be significantly more dedicated to the use of commercial software and less interested in exploratory programming, seemed to indicate that they had a greater need for practical applications from their computing than did the Dependents.

If the above supposition was true, one would have surmised that the Owners would have produced more useful end-products from their computing than the Dependents, but no significant differences were found. This can be explained, however, when one examines the equipment owned and the methods used by the Owners. Practical applications are not always easy to achieve on a small computer, and this is especially true if the computer is used as it stands, without adaptation and expansion, and if the programmer is only familiar with BASIC (as were most of the Owners). This high-level language requires a larger memory capacity than programs written in machine code and is much slower in its performance. Although one can purchase software written in machine code, this is frequently very expensive and it is not unusual for business software to cost many hundreds of pounds, far more than the purchase price of many computers. Without the skills necessary to copy and pirate such software, the purchase of these programs becomes prohibitive to many microcomputer owners, leaving only the cheap games software within their budgets. One may of course write one's own programs in machine code, but this requires a high degree of dedication and time for successful use, and the Owners were shown not to make these investments. Such factors could have accounted for the Owners' lack of useful end-products from their computing, in spite of their apparent practical attitude towards programming.

Similarly, many of the Owners, in spite of their initial financial outlay and interest, now rarely used their home computers, with eleven reporting that they were never used at the time of this survey. It could be true to say that even when in possession of a home computer there is very little which can be performed which is of any practical value to the majority of the population. Most adults do not keep household accounts, and the use of spreadsheets to keep track of what normally would be worked out on the back of an envelope would be seen by many as an unnecessary expenditure of time and effort. Even word-processing, a very valuable use of a microcomputer, is of little use to most people unless one is an avid letter-writer with typing skills or running a business. However, even these relatively straightforward tasks demand the use of expensive software, disc-drives and hardcopy printers, together with the patience to work through the often lengthy manuals before one can achieve the most rudimentary of results. No doubt many home computer owners became reluctant to make further such investments for such infrequently used benefits, (although this situation may have changed since 1987 when dedicated word-processors became available at much lower prices). Although the microcomputer appeared to promise a cornucopia of benefits these aspirations were rarely fulfilled during the early 1980s, and for this reason the interests of most computer owners, including the group of Owners, were therefore unlikely to have been satisfied if practical applications were desired.

Although one could assume that the Dependents, with their mass of hardware and software, could have been using their home computers for extrinsically worthwhile purposes, this was not found to be so. Although the potential was there, the majority of the Dependents still concentrated their activities upon exploration of the system, using the computer as a vehicle to experiment with the new technology. This activity was still being enjoyed even after many years of computer ownership, by which time one would imagine that such interest would pall. Although a small number of the Dependents did state that they had purchased a computer with the intention of carrying out practical applications, few of them fulfilled such aims.

> 'I bought it for a possible business. I played a lot more games than I thought I expected to and wrote fewer programs. I didn't write anything saleable. I think I had some grand illusions about hooking up the central heating—in less realistic moments they still persist!'

> 'It was a natural development from my interest in electronics. I got it to control household functions—I was very naïve in those days. I actually spend my time exploring and playing with the system.'

The majority of the group made no such claims. Most had bought their computers in order to satisfy their curiosity about new technology and for the sole purpose of learning more about computers and programming, not for performing specific functions. Although they wrote programs, most were rarely completed. The process of programming proved more important than the end-product.

> 'I've never really put it to any practical purpose. The satisfaction is setting a problem and getting it to do it.'

> 'It's for sheer intellectual satisfaction—the feeling of doing a job well—it's extremely soothing. You can see whether an algorithm will work. It doesn't matter if you waste time at home, you don't have to account for it except to your wife.'

> 'I've started and never finished a big program, but I feel I could write anything I wanted to in any language you could name.'

The difference between the Dependents and the Owners may be seen as similar to that between a pure and an applied scientist, or between two types of car owners; the first of whom wishes to dismantle and continually to re-tune the engine and the second who is more concerned with getting from A to B but understands sufficient to carry out basic maintenance.

> 'I do it mainly as a hobby. There are no really useful end-products except for me. It's a fairly selfish occupation. I program to practise writing programs.'

> 'It's a bit like marking your own exam papers; correcting mistakes. It's so satisfying because you know there is a right answer, it's like building a complicated bit of machinery.'

> 'It's straight hands-on, no flow charts. I'm untrained and inefficient, but I do brilliant things in an inefficient manner.'

Programming languages were learnt, many of which the Dependents stated they had no real need of or applications for, and their additional interest in the now almost archaic low-level languages confirmed the depth of their learning needs. Low-level languages written in machine code and assembler are extremely time-consuming to compose and demand a high degree of accuracy if they are to run. However, they do allow one more fully to understand the workings of the computer as one is programming directly in the computer's own code. The Dependents' occasional use of compilers for converting high-level languages to machine code did not obviate their desire to learn machine code first-hand, again indicating that efficiency was of less importance than learning.

The Dependents' need for first-hand personal experience at the expense of speed and efficiency was also confirmed by their self-tuition in languages and by their reluctance to pre-plan their programs, in contrast to the Owners' methods. The 'hands-on' programming style favoured by most of them increased the likelihood of program failure and the need to debug the code. Although the interviewees stated that the basic structure of their programs was often written in a few hours, many months were often spent debugging them; the refining, trouble-shooting, fault-finding activities which for many brought the greatest stimulation.

> 'I spend most of my time debugging, it's the most enjoyable part. I spend a lot of my time thinking but I don't work things out the hard way. I sit at the keyboard and it develops and grows.'

> 'It's the greatest part of the overall challenge. If I knew exactly what was going to happen it would be boring—that's when it (the computer) becomes just a tool.'

The thought of using the computer as merely a tool was anathema to many of the Dependents; people using the computer in this way were seen to be limiting its scope and potential to that of a machine, rather than as an intellectual and technological marvel designed by a genius. Although full of admiration for computer innovators, (Sinclair was described by one interviewee as the 'high priest), the Dependent interviewees revealed that much of their exploratory computing was undertaken to test their abilities against those of the professionals, with 'hacking' into the computer systems and databases of others a favourite method. Even if outwardly dishonest, few appeared to disapprove of this practice, an attitude shared by the Owners. It was therefore not higher moral principles which prevented the Owners from hacking, but lack of exploratory interest and expertise, and to an extent lack of ownership of modems.

The need to challenge and understand the work of others was also reflected in the Dependents' need to break the supposedly impregnable protection codes of commercial software, using techniques which were often complicated and lengthy. They enjoyed the challenge that code-breaking presented to their computing skills and once in the program they felt free to explore and adapt it at will. One stated that it was 'an honourable part of being a junkie—to challenge and beat the professionals'.

Collectors of 'pirated' software who were interviewed appeared to be not unlike stamp-collectors, except that there was little financial profit to be made in this instance. Elaborate systems of cataloguing were designed, and programs were exchanged with others of like mind. There was intrinsic pleasure to be gained, and many possessed large quantitites of utilities, business and communications software, word-processing packages and games, most of which were never used. By contrast very few of the Owners pirated software or adapted it to their own needs, once more indicating a lack of exploratory ability and interest in developing the level of skill required; not so the Dependents:

> 'I just want to explore them [software programs]. They are usually adventure games and new arcade games, but I get bored when I've mastered them. They are intellectually challenging, I like using my brain to solve things. I play all the new good games once and I try to expand them.'

As so many differences were observed between the two groups it had also been anticipated that the Dependents would use the networks for different purposes than did the modem-owning Owners. It was anticipated that the Dependents would concentrate upon gaining computing information and would be less likely to use the facilities for personal communication with others; however no differences were observed. Communication via networks remains relatively impersonal and anonymous, and could be seen to provide the Dependents with regular contact with people with similar interests to themselves while not demanding the degree of intimacy necessary in face-to-face encounters.

One may conclude from these results that although there were some similarities between the computing activities of Dependents and the Owners, which may be characteristic of a male attitude to new technology, there were sufficient significant results to show differences not only quantitatively but also qualitatively. Although some of the Owners were inhibited from undertaking certain activities by their lack of peripheral equipment, this lack was one of the most significant indicators of their lower interest levels in computers in general. There was no reason to suggest that the Dependents were more affluent or had more spare time than the Owners, but the results clearly showed that the Dependents were more likely to be using the computer as a means of learning about the technology and as a means of stretching their own intellectual, and perhaps playful, interests. They were prepared to devote both time and money to an interest which had for the most part been longstanding, while the Owners confined themselves to a few hours a week, using commercial software rather than themselves programming or exploring the system.

Using the computer as a means of learning about technology, by experimenting with its capabilities, rather than for task-specific goals could be said to distinguish the computing activities of the Dependents from those of the Owners. Nevertheless the results from the interviews demonstrated that not all of the Dependents were interested in the same types of computing activities. The interviewees each appeared to have their own speciality, an activity upon which they concentrated most of their time and energy. Some of the Dependents rarely if ever carried out any programming, preferring to devote their time to networking, while others became expert hackers.

'Compulsive programming' (Weizenbaum, 1976) was not therefore a universal activity among the group, although it did appear to have been the major activity upon purchase of the computer. Initial impressions gained during the interviews indicated that certain personality types seemed to be associated with each type of activity and that certain needs were being met by them. Although the majority of this research concentrates on the variables upon which the Dependents differed from the control groups, further investigations were carried out to discover if the perceived types of computer dependency were able to be isolated and measured. (Discussion of the different types of computer dependency are given in Chapter 9).

Conclusions

These results confirmed that there was great concordance between the computing activities of the students in the preliminary study and the group of Dependents, and that these were compatible with some of the descriptions found in the reviewed literature (Chapter 2). The Dependents' interests were centred upon exploration of the computer system, by investigative and innovative programming, by breaking into and adapting the programs of the professionals, and by learning about the hardware and software by direct, hands-on experience. The computer was often viewed as a toy and the Dependents were not primarily interested in producing workable programs, although many had been working on lengthy projects which some thought they would one day complete. Their aim was self-education, gained by rising to the intellectual challenge presented when using computers, the satisfaction of curiosity and the increase in knowledge about a widely-used but little understood modern artefact. Computing had become dominant in their lives, leaving them little time or inclination to undertake any other activities.

Although both the Dependents and the Owners had shown sufficient interest to stimulate purchase of a home computer, their initial experiences and beliefs had led them to approach the computer with differing attitudes and expectations. While the Owners, with their professionally-trained skills, appeared to wish to use the computer as a functional, problem-solving tool (an expectation which appeared not always to be satisfied adequately), the needs of the self-taught Dependents were not frustrated by the practical limitations of the micro-computer. They had purchased their machines in order to find out about computers, and not only were they sufficiently motivated to master new technology by experimentation, they were also prepared to invest heavily in both time and money. Inertia and boredom did not occur, as the process of programming was the main function of their interaction, in order to study new technology. If one machine was 'mastered', another could be purchased. Some of the interviewees had spent over five years exploring the computer's potential and still felt there were many 'thousands of man-years' interest left in this activity. As the hardware and software became more sophisticated, so they continued to expand their knowledge and expertise to match the technology.

From the questionnaire results and the interviews with the Dependents, together with discussions with less enamoured computer users, it was possible to see that the features which held the Dependents' enthusiasm were the same ones which caused frustration for other users. This difference mirrored that found by Malone (1984) when analyzing which features of computer games made them fun to play. He concluded that the 'requirement for good toys and good tools are mostly opposite'.

One's attitudes and needs determine the system which is required, and although the Dependents saw computers as intrinsically interesting, others, who wish merely to use them as tools, tend to see them as sources of conflict rather than pleasure. Most users neither wish nor have the time to explore the computer at will, especially while at work, and do not rise to the challenges presented. Experimental trial-and-error methods tend to reduce efficiency and the likelihood of accuracy, hence for most people there is a need to design straightforward computer systems which have good 'help' facilities and clear, concise and well-indexed manuals. However, unlike task-centred users of tools, the Dependents did not want the computer to be a 'black box' which could be used with little effort; they delighted in its intricacy, and were mainly unconcerned with practical end-products. They revelled in the jargon and the fact that risk and experimentation had to be entertained in order to exploit computers to the full, they were unconcerned by the poorly written manuals, and enjoyed having to learn a 'foreign' language in order to use the machine. They had no desire to use an artefact which was 'transparent' in its simplicity, as time and efficiency were of little importance to them. Therefore, although it was possible to isolate the features of the computer and computing which had encouraged the Dependents to view the machine as a fascinating toy, such information would be of little use to the computer industry if desirous to redesign their systems in order to encourage the frustrated or hostile user to view them more favourably.

These results established that the Dependents' needs for self-education and enjoyment when using computers differentiated them from other computer users. In order to discover if such needs had been apparent before the computer had entered their lives, further research was undertaken to see if their previous hobbies also fulfilled similar functions, and to discover what factors within the personalities of the Dependents had been influential in the development of these interests.

FOOTNOTE The investigation into computing activities concentrated only upon the responses from the two computer-owning groups, the Dependents and the Owners, specifically because the interviewees stated that the activities undertaken on their home computers reflected their true interests. However, the Non-owners were also asked to describe their computing experiences, and the results showed that 70 of the 75 Non-owners had also used computers, usually at work, while studying, or at a friend's home (confirming the difficulty experienced when trying to find computer-naïve people who matched the Dependents for age, sex and educational levels).

Fifty-two of the Non-owners had attempted programming, 31 of whom had enjoyed it, but most of their time was spent using commercial software or games-playing. This group also tended to hold very positive attitudes towards new technology, as can be seen in the frequency of their responses to the question asking for their level of interest in computers.

4—no interest 10—mild interest 26—interested
26—very interested 9—extremely interested

Although they did not own computers and were therefore unqualified to answer many of the questions included within this chapter, as such a large proportion of the Non-owners had been computer users their data were included within the remaining comparative investigations carried out between the groups.

Chapter 7
Computing and other leisure activities: Was computing an isolated activity?

Introduction

The interviews with the five undergraduate students carried out at the beginning of this research project suggested that their interest in computing did not occur in isolation but appeared to be a natural progression from their previous hobbies. All had been interested in electronics and mechanical activities, seemingly from an early age, and much of their interest in computers stemmed not only from their readings of non-fiction but also from science fiction. All five were also interested in various logic games and mental puzzles which could be likened to the investigative, exploratory nature of their computing activities. In addition, none spent much time socializing and many stated that they disliked most sporting activities, especially team sports. All were very interested in music, and as music and mathematics show structural fidelity it was possible to see that the pattern of their previous activities could be related to their current interest in computers. Not only this, but the students stated that they tended to be the types of people who were not eclectic in their hobbies but preferred to concentrate upon just a few activities which they took very seriously and studied assiduously.

As the range of hobbies listed by the students appeared to show more than coincidental similarity it was thought that a profile of particular past hobbies could have been a necessary precursor for an interest in computers to develop to the levels of ownership and dependency. Further research was therefore initiated to investigate the following hypotheses:

(1) If computer dependency develops in those with a history of a few hobbies studied to an intense level, differences in levels of interest in hobbies and in the numbers of activities attempted should be observable between the Dependents and the control groups.

(2) If computer dependency is influenced by a profile of particular hobbies, differences would be observed between the Dependents and the control groups in their choice of favourite activities.

(3) From this, computer ownership itself is also likely to have developed from a particular set of hobbies; therefore it was likely that the Dependents and the Owners would share a particular set of hobbies which would be of little interest to the Non-owners.

(4) If no differences were found between the three groups for certain hobbies, such activities could be seen as representative of people showing similar demographic characteristics to those of the three matched groups.

Using the computer in the manner favoured by the Dependents seemed unlikely to have developed in a vacuum. Perhaps their desire for intellectual understanding and control over artefacts had been apparent throughout their lives, with the computer merely acting as the most recent in a series of intense activities which satisfied those needs. If significant differences were found between the hobby profiles of these three groups, the results could go some way to explain why some of the people who develop an interest sufficient to warrant purchase of a computer become dependent upon it.

A comparative investigation of leisure activities

Methods of investigation

The methods used to investigate this area took three forms. Firstly, an open-ended question was included in the General Questionnaire which asked the Dependents to list any past hobbies which had been of especial interest to them, in order to discover if there were similarities between their responses. Secondly, comparisons were made between the results from the Dependents and the control groups to further questions (included in the General Questionnaire) which examined the various types of interests and hobbies of the groups. The third and major part of this study involved the use of an extensive inventory of hobbies and activities. This was devised in order that rigorous statistical analysis could be applied to the data, and from which a global view of the hobbies of the three groups could be fully examined. Interview data also supplemented the results obtained.

It had been anticipated that an existing inventory could be used effectively, but study of the literature revealed that most were centred upon vocational interests rather than leisure activities.

None were at all suitable for this study; most were far from comprehensive and many, because of their age, lacked modern activities such as electronics, squash or even television watching. The majority lacked relevance for groups of adult males and did not include many of the activities which had already been shown to be of great interest to the Dependent computer owners. Therefore an inventory specific to this research was devised, which included activities pertinent to the ages and interests of the groups examined; the *Leisure Activities Inventory*. The aim of the inventory was to deduce which of the activities had been attempted by each person and to discover their level of interest in that activity. The choice to undertake an activity is an expression of the initial level of interest, and continuing interest gives a measure of whether the activity fulfils the desires of the individual.

The inventory included many of the hobbies mentioned by the Dependents, either on the General Questionnaire or during the initial interviews, but more were added. Activities were included which could logically be termed hobbies, but also others such as 'dining-out' and 'holidays' which would perhaps not be thought of as such. Numerous activities of a social and sporting nature were also included in the inventory, with the aim of determining whether the group of Dependents were as uninterested in them as were the interviewed students. Reading was divided into sections in order to isolate science fiction from other fictional literature, and such items as religion, politics and voluntary work were also included in order to give as good a spread of activities as possible. Various aspects of computing were included in order to investigate just which activities had been attempted and which were of greatest interest to all three groups. Altogether 128 activities were listed.

The instructions which accompanied the Leisure Activities Inventory stressed that only those activities which had been undertaken voluntarily in their spare time were relevant to this study, not those which may only have been undertaken at school, for example, where choice is often greatly restricted. The design also allowed respondents to state that they had undertaken certain activities but had no interest in them.

A numerical scoring mechanism was employed for the interest level ratings: 0 indicated 'No interest'; 1, 'Mild Interest'; 2, 'Interested'; 3, 'Very Interested'; and 4, 'Extremely Interested'. If the hobby had never been attempted the subjects were asked to delete all of the codes.

In order to present clear, unambiguous results from such a large quantity of data a categorization scheme was needed for the list of the activities showing differences between the groups. Other researchers using inventories for the examination of leisure activities had tended to use factor analysis for this purpose, although often dealing with missing data in their results. However, this form of analysis seems prone to yield bizarre results, as was shown by Fiction and Outdoor Sporting coming under the factor 'Ideas and Fancy' (Thorndike, 1935); Camping under 'Athletics', and House Repairs, Listening to Music and Discussion all under the factor labelled 'Domestic' (Vernon, 1949). Guildford *et al.* (1954) listed Amusement and Humour under 'Physical Fitness'; McKechnie (1974) placed Wrestling under 'Mechanics'; and Nias (1975) included music under 'Science', Table Tennis under 'Pop Music' and amusingly, Going on

and the arts. However, only the Dependents considered that their interest in these four major subject areas had increased since using computers. It was also confirmed that the majority of the Dependents considered computing to be their best hobby; a sentiment shared by very few of the Owners.

Although anticipating that the Dependents would have experienced more changes to their social lives than the Owners since purchasing a computer, few of either group reported any difference. This result was clarified however by the responses to a question asking all three groups to specify the number of evenings spent socializing. The Dependents were shown to socialize far less than the other two groups, with the median for the Dependents at one evening per week and for the control groups three. This result correlated with the comments from the interviewed students who stated that they had never led full social lives and who, if computing had not become of interest, would choose to spend their evenings undertaking other hobbies rather than social activities. Although more Dependents than Owners were members of computer clubs, this only applied to one third of them, indicating that even this method of forming relationships with like-minded people was rarely undertaken.

As mentioned previously, both computer groups were primarily interested in scientific and technical subjects, while significantly more Non-owners were interested in the arts. A significant trend was also observed between the three groups' reported interest in sport while young, with the Non-owners showing the greatest interest and the Dependents the least. All stated that they read a great deal, and while young and currently the Non-owners read more fiction than either the Dependents or the Owners, and as adults the Dependents read more science fiction than did the controls.

These results demonstrated that differences in interests did exist between the three groups, and suggested that each exhibited an individual profile of hobbies. The application of the Leisure Activities Inventory was able to add further confirmation and clarity to these results.

The Leisure Activities Inventory

Before examining each hobby on the Leisure Activities Inventory individually, initial analyses were carried out upon the total set of activities attempted by each group to determine whether the Dependents were more likely than the controls to practise fewer hobbies at a more intense level. The results showed that the average number of activities which had been attempted by each group was approximately 50 per cent of the total number possible and no significant difference was observed. Further analysis showed that there was also no significant difference between the groups for the total frequencies of interest level shown, with 'Mild Interest' being the level most frequently allocated by all groups (see Table 7.3, Analyses 1 and 2).

Table 7.3 Frequencies of responses for total set of leisure activities

	Dependents		Owners		Non-owners	
	Total	Means	Total	Means	Total	Means
Numbers of activities attempted Analysis 1 ($p > 0 \cdot 10$)						
Attempted	5067	67·6	5162	68·8	4949	66·0
Not attempted	4533	60·4	4438	59·2	4651	62·0
Totals	9600	128·0	9600	128·0	9600	128·0
Interest levels in activities attempted Analysis 2 ($p > 0 \cdot 10$)						
No interest	1205	16·1	1040	13·9	849	11·3
Mild interest	1467	19·6	1576	21·0	1527	20·4
Interested	1160	15·5	1288	17·2	1290	17·2
Very interested	669	8·9	736	9·8	814	10·8
Extremely interested	566	7·5	522	6·9	469	6·3
Totals	5067	67·6	5162	68·8	4949	66·0
Interest levels in activities attempted Analysis 3 ($p < 0 \cdot 01$)						
No/extremely interested	1771	23·6	1562	20·8	1318	17·6
Other levels of interest	3296	44·0	3600	48·0	3631	48·4
Totals	5067	67·6	5162	68·8	4949	66·0

However, further observation of the data appeared to indicate that the Dependents were more likely than the other groups to state that they had 'No interest' in many of the activities they had attempted, as well as being more likely to use the other extreme category of interest level, 'Extremely interested', for the hobbies they did enjoy. When the data were re-grouped and re-analyzed (Analysis 3), it was shown that the Dependents were significantly more likely to use the extreme categories than were the control groups. This suggested that although the Dependents had made equivalent numbers of attempts at the activities, they rejected more as being uninteresting than did the controls and concentrated on fewer hobbies with a higher level of interest. The results also showed a significant trend between the three groups, with the Non-owners using the intermediate categories more frequently than either of the other groups and the Dependents using them the least.

Individual leisure activities showing no significant differences

When the individual activities were analyzed, 41 of the 128 hobbies showed no significant differences between the three groups, either in the number who had attempted that activity or in their reported interest levels. The 41 activities are given in Table 7.4. These are sub-divided according to the numbers of people who had attempted the activity, as it was felt that many had yielded non-significant results either because of their rarity or popularity. 'Rare Activities' had been attempted by less than one third of the respondents, 'Universal Activities' by at least two thirds, while 'Other Activities' included the mid-range.

Table 7.4 Leisure activities showing no significant differences

Rare Activities	Universal Activities	Other Activities
Archery	Board games	Bridge
Bellringing	Cinema going	Collecting, other
Fencing	Cooking	Drawing/painting
Flying	Crossword puzzles	Extra education
Gliding	Extra work own time	Gambling
Knitting	Jigsaw puzzles	Learning language
Martial arts	Music, listening to	Motor cycles
Meteorology	Reading, general	Photo. processing
Music composing	Reading, fiction	Pornography
Parachuting	Reading, non-fiction	Religious activity
Pets, breeding	Television watching	Shooting
Pottery/sculpture		Singing
Racing, greyhound		Voluntary work
Racing, pigeon		Wine/beer making
Scuba diving		
Skiing		
Video/cine filming		
Yoga		

These activities, yielding no significant differences between the three groups, can be seen as representative of people exhibiting the same demographic characteristics and are no longer relevant to this study.

Individual leisure activities showing significant differences

Significant differences were observed between the groups for two-thirds (85 out of 128) of the leisure activities listed, either for differences in the numbers attempting the activity, in interest level, or in both. These yielded a total of 130 significant results, 76 of which showed differences in interest levels between the groups.

The results seemed to suggest that there were great differences in the hobbies of interest to the Dependents, the Owners and the Non-owners, with the two groups of computer owners sharing some interests, the two control groups sharing others, but with few being shared by the Dependents and the Non-owners.

To determine how the hobbies of the groups were correlated, the set of hobbies which showed significant differences in interest level between the groups were subjected to a correlation analysis (Spearman rank correlation coefficient) carried out on each pair of groups, i.e. the Dependents and the Owners, the Owners and the Non-owners, the Dependents and the Non-owners. In order to makes this assessment the number within each group reporting themselves as either 'Very interested' or 'Extremely interested' in each of these activities was determined and the rankings in popularity of the total set of the 62 hobbies were used in the analyses. The results were as follows:

Dependents: Owners	rho=0·603	z=4·713	p<0·001
Owners: Non-owners	rho=0·729	z=5·692	p<0·001
Dependents: Non-owners	rho=0·267	z=2·085	p<0·025

The results were in line with the original hypotheses and showed that there was a much stronger correlation of interest in leisure activities between the two groups of computer owners and between the two control groups than between the Dependents and the Non-owners. The results from the individual hobbies confirmed these initial generalizations.

The results from the individual activities are given in Tables 7.5 to 7.11, one for each of the seven categories of hobbies, and are divided under sub-headings according to where differences were observed between the three groups. The sub-headings are worded to indicate the trend of the attempts and interest levels for each activity between the groups; for example, under the heading 'Dependents less than (Owners and Non-owners)' the Dependents were shown to have attempted Cricket less and were less interested in this activity than the two control groups.

Each table shows the significance levels of the frequency of respondents attempting the activity and/or their levels of interest in it, as follows: $^{*}p<0{\cdot}05$, $^{**}p<0{\cdot}01$, $^{***}p<0{\cdot}001$. Although most of the significant results showed differences between one group and the other two, on a few occasions differences were observed between all three groups and in these instances the activity is listed under two separate headings within the table, to illustrate where each difference lay.

The results from the *Sporting Activities* proved quite conclusively that in spite of the wide variety of activities included none showed the Dependents attempting them or interested in them more than either of the control groups (see Table 7.5). The Dependents were less likely than the controls even to have attempted the activities, and both the Dependents and the Owners showed less interest than the Non-owners in many of activities they had experienced.

Similarly, the Dependents were less likely than the controls either to have attempted or to show interest in the *Social Activities*. In two instances, those of entertaining and dining-out, neither the Dependents nor the Owners were as interested as the Non-owners, and they were less likely to go dancing, but for the most part the results showed the two control groups to have equivalent responses (see Table 7.6).

By comparison the results for the activities labelled *Games and Puzzles* and *Computing and Electronics* showed reverse trends from those above (see Tables 7.7 and 7.8). Although the Dependents and the Owners were more likely than the Non-owners to have attempted some of the Games and Puzzles, where differences in interest occurred these invariably showed the Dependents with the highest levels of interest. For the

Table 7.5 Differences in attempts and interest levels for the Sporting activities. n.s., no significant differences

No.	Sporting activities	Difference in attempts	Difference in interest
Dependents more than (Owners and Non-owners)		n.s.	
(Dependents and Owners) more than Non-owners		n.s.	
Dependents less than (Owners and Non-owners)			
8	Athletics	**	**
9	Badminton	**	
16	Car racing, spectator	*	
32	Cricket	*	***
34	Cycling		***
49	Football	**	***
53	Go-Karting		*
54	Golf	*	***
57	Hiking/walking	**	**
61	Jogging	**	
88	Potholing/caving	**	
94	Racing, horse	*	**
104	Rugby, playing		*
105	Running	*	
112	Squash	***	
113	Swimming		*
115	Tennis	**	**
122	Watching sport	***	***
124	Windsurfing	***	
(Dependents and Owners) less than Non-owners			
8	Athletics		***
21	Climbing	**	*
34	Cycling	*	
47	Fishing	**	
59	Horse riding		**
61	Jogging		**
62	Keep fit	**	***
105	Running		*
106	Sailing	**	
113	Swimming		*
115	Tennis	**	
123	Weight training		*
Owners more than (Dependents and Non-owners)			
106	Sailing		**

activities categorized as *Computing and Electronics*, in many instances there were differences between all three groups for the numbers attempting the activities, with the Dependents showing the most and the Non-owners the least attempts. Once again the Dependents showed higher interest levels in many of these activities than did either of the control groups, but both the Dependents and the Non-owners of computers showed higher interest in computer music than did the Owners.

Table 7.6 Differences in attempts and interest levels for the Social activities

No.	Social Activities	Difference in attempts	Difference in interest
Dependents more than (Owners and Non-owners)		n.s.	
(Dependents and Owners) more than Non-owners		n.s.	
Dependents less than (Owners and Non-owners)			
19	Children, being with	⋆	
36	Darts		⋆
37	Dining out		⋆⋆
38	Discotheques		⋆
80	Party-going		⋆⋆⋆
90	Pubs, visiting	⋆	⋆⋆⋆
111	Snooker	⋆	
(Dependents and Owners) less than Non-owners			
35	Dancing	⋆	
37	Dining out		⋆⋆
43	Entertaining		⋆⋆⋆

Table 7.7 Differences in attempts and interest levels for the Games and Puzzles activities

No.	Games and puzzles	Difference in attempts	Difference in interest
Dependents more than (Owners and Non-owners)			
2	Adventure games		⋆
5	Arcade games		⋆⋆
17	Card games	⋆	
18	Chess	⋆	
41	Dungeons and Dragons		⋆
66	Maths/logic puzzles	⋆⋆⋆	
121	War gaming	⋆⋆	⋆
(Dependents and Owners) more than Non-owners			
2	Adventure games	⋆⋆⋆	
5	Arcade games	⋆⋆⋆	
41	Dungeons and Dragons	⋆⋆⋆	
66	Maths/logic puzzles		⋆
Dependents less than (Owners and Non-owners)		n.s.	
(Dependents and Owners) less than Non-owners		n.s.	

were not so clearly defined and showed a variety of differences between the groups. Within *Practical Crafts* (Table 7.9) the two groups of computer owners were more likely to have undertaken hobbies concerned with models and model-making but were less interested in car maintenance; while the two control groups showed more interest in Do-It-Yourself and gardening than did the Dependents.

Within *The Arts* category (Table 7.10), the Dependents were shown to be significantly more interested in science fiction than the control groups, as had been anticipated, but less so in playing musical instruments or theatre-going. While the Non-owners were more interested in poetry

Table 7.8 Differences in attempts and interest levels for the Computing and Electronics activities

No.	Computing & Electronics	Difference in attempts	Difference in interest
Dependents more than (Owners and Non-owners)			
24	Computer chess	★★★	
25	Computer games	★★★	★★★
26	Computer graphics	★★★	★★★
27	Computer hacking	★★★	★★★
28	Computer music	★★★	
29	Computer networking	★★★	★★★
30	Computer programming		★★★
42	Electronics	★★★	★★★
91	Radio, amateur	★	
(Dependents and Owners) more than Non-owners			
24	Computer chess	★★★	
25	Computer games	★★★	
26	Computer graphics	★★★	
27	Computer hacking	★★★	
28	Computer music	★★★	
29	Computer networking	★★★	★★★
30	Computer programming	★★★	★★★
42	Electronics	★★★	
56	HiFi technology		★
91	Radio, amateur	★	
92	Radio, CB	★★	
Dependents less than (Owners and Non-owners)			
		n.s.	
(Dependents and Owners) less than Non-owners			
		n.s.	
(Dependents and Non-Owners) more than Owners			
28	Computer Music		★

than the computer owners, both the Non-owners and the Dependents had attempted amateur dramatics more often, and showed greater interest in philosophy and writing than the Owners.

The remaining activities which were unable to be classified, the *Miscellaneous* activities (Table 7.11), also showed a variety of differences between the three groups, but the majority of these results showed the differences to lie between the Dependents and the two control groups.

In order to determine which of the activities were of greatest interest to each of the three groups, the first 25 activities which had been rated most frequently under the categories 'Very interested' or 'Extremely interested' for each group were determined. These are shown in Table 7.12, in descending rank order of interest. (The frequency of those who used the codes 'Very interested' or 'Extremely interested' are given under the heading Frequency, with a maximum possible of 75.) Forty different hobbies were represented in total, and a few of the activities which have

Table 7.9 Differences in attempts and interest levels for the Practical Craft activities

No.	Practical Crafts	Difference in attempts	Difference in interest
Dependents more than (Owners and Non-owners)			
78	Needlework		**
(Dependents and Owners) more than Non-owners			
70	Model making	*	
71	Model railways	***	
93	Radio controlled models	*	
Dependents less than (Owners and Non-owners)			
39	DIY		**
51	Gardening		**
55	Handicrafts	*	
84	Photography		*
127	Woodwork		**
(Dependents and Owners) less than Non-owners			
15	Car maintenance		**
Owners more than (Dependents and Non-owners)			
68	Metalwork	*	
127	Woodwork	*	

Table 7.10 Differences in attempts and interest levels for the Arts activities

No.	Arts activities	Difference in attempts	Difference in interest
Dependents more than (Owners and Non-owners)			
102	Reading, science fiction		***
(Dependents and Owners) more than Non-owners		n.s.	
Dependents less than (Owners and Non-owners)			
75	Music, playing instrument		*
116	Theatre-going		**
(Dependents and Owners) less than Non-owners			
101	Reading, poetry		*
(Dependents and Non-owners) more than Owners			
3	Amateur dramatics	*	
83	Philosophy		*
126	Writing stories/articles		*

been classed as 'Universal' in Table 7.3 were also included because of their popularity.

For additional clarity the sets of hobbies which were of greatest interest are also redisplayed in Table 7.13, this time divided according to the seven categories of activities. This shows that there was on occasion total omission of hobbies for some groups under certain of the headings, and also that some activities were popular with all three groups.

Observation of the rank orders in Table 7.12 seemed to suggest that there was almost a reversal of favourite hobbies between those of the Dependents and the Non-owners, as was confirmed by correlation

Table 7.11 Differences in attempts and interest levels for the Miscellaneous activities

No.	Miscellaneous	Difference in attempts	Difference in interest
Dependents more than (Owners and Non-owners)			
7	Astronomy	**	**
77	Mythology	*	
(Dependents and Owners) more than Non-owners			
1	Acquaria	**	
97	Railways, steam	*	
Dependents less than (Owners and Non-owners)			
1	Acquaria		**
4	Antiques	*	***
14	Camping		***
58	Holidays		***
82	Pets, keeping		*
86	Politics, active in	*	
(Dependents and Owners) less than Non-owners			
11	Bird watching		*
(Dependents and Non-owners) more than Owners			
22	Collecting stamps		*
67	Meditation		*
72	Model soldiers		*
Owners more than (Dependents and Non-owners)			
117	Train spotting	*	
118	Travelling/sightseeing		**

analysis between the pairs of groups (Spearman rank correlation coefficients). The results were as follows:

Dependents: Owners	rho = 0·211	z=1·317	p>0·05
Owners: Non-owners	rho = 0·723	z=4·512	p>0·001
Dependents: Non-owners	rho = − 0·171	z=1·069	p>0·05

The results showed a low correlation between the preferred hobbies of the Dependents and the Owners, a high correlation between the Owners and the Non-owners, and a negative correlation between the Dependents and the Non-owners. These results were confirmed by the fact that of the 130 significant differences which were observed between the groups, 76 were found to lie between the Dependents and the two control groups, 42 between the Non-owners and the two groups of computer owners, and only 12 showed the responses of the Dependents and the Non-owners to be similar but different from those of the Owners.

The results obtained from the Leisure Activities Inventory had therefore been able to confirm the hypotheses. There were hobbies which were of interest only to the Dependents and the Owners, which differentiated between the interests of the two groups of computer owners and the Non-owners. In addition, the results showed that it was possible to isolate

Table 7.12 Ranked activities of most interest to each group, from ratings for the Very and Extremely Interested categories

Dependents			Owners			Non-owners		
Rank	Activity	Freq	Rank	Activity	Freq	Rank	Activity	Freq
1	Comp. program	56	1	Holidays	42	1	Holidays	51
2	Comp. networks	52	1	Music, listening	42	2	Music, listening	49
3	Comp. graphics	39	3	Travelling	35	3	Dining out	39
3	Science fiction	39	4	DIY	29	4	Reading, general	35
5	Comp. hacking	37	4	Reading, general	29	5	Fiction	33
6	Non-fiction	36	6	Entertaining	28	6	Pubs, visiting	29
7	Music, listening	35	6	Hiking/walking	28	6	Non-fiction	29
8	Comp. games	29	6	Non-fiction	28	6	Swimming	29
8	Reading, general	29	9	Comp. program	27	9	Camping	27
10	Electronics	27	9	Photography	27	10	Cooking	26
11	Fiction	25	9	Pubs, visiting	27	10	DIY	26
12	Holidays	23	12	Dining out	25	10	Entertaining	26
13	Comp. music	22	12	Fiction	25	10	Hiking/walking	26
13	Dining out	22	14	Science fiction	23	14	Travelling	25
15	Adventure games	20	15	Cooking	22	15	Cinema-going	24
16	Photography	19	15	Swimming	22	15	Keeping Fit	24
16	TV watching	19	17	Party-going	20	17	Pets, keeping	20
18	Children, with	18	18	Comp. graphics	19	17	Theatre-going	20
19	Cinema-going	17	19	Cinema-going	18	19	Car maintenance	18
19	Cooking	17	20	Children, with	17	19	Party-going	18
19	HiFi technology	17	20	Gardening	17	19	Watching sport	18
22	DIY	16	20	HiFi technology	17	22	Children, with	17
22	Maths/logic puzz	16	20	Pets, keeping	17	22	Jogging	17
24	Arcade games	15	24	Cycling	16	24	Cycling	16
24	Astronomy	15	25	TV watching	16	24	TV watching	16

those activities which were of interest to or were avoided by those who became dependent upon computers. Each group was therefore interested in different types of hobbies and activities, and from the results it was possible to assemble a composite profile of activities enjoyed or avoided by each of the three groups. Even some of the nine results which showed the Owners results as different from those of the Dependents and the Non-owners, results which were not accounted for by the hypotheses, could also be sensibly incorporated into the profiles for the groups.

Hobby profile of the Non-owners

The hobbies of interest to the Non-owners showed the greatest variety and diversity of all three groups. Earlier results showed that while the Dependents and the Owners were mainly interested in scientific and technological subjects the Non-owners were as equally interested in the Arts as in these areas, as was reflected by their favourite hobbies.

The Non-owners were interested in all manner of social activities,

Table 7.13 Activities of most interest to each group, by type

Dependents		Owners		Non-owners	
Rank	Activity	Rank	Activity	Rank	Activity
Sporting activities					
		6	Hiking/walking	6	Swimming
–		15	Swimming	10	Hiking/walking
		24	Cycling	15	Keep Fit
				19	Watching sport
				22	Jogging
				24	Cycling
Social activities					
13	Dining out	6	Entertaining	3	Dining out
18	Children, with	9	Pubs, visiting	6	Pubs, visiting
		12	Dining out	10	Entertaining
		17	Party-going	19	Party-going
		20	Children, with	22	Children, with
Computers/electronics					
1	Comp.. programming	9	Comp. programming	–	
2	Comp. networks	18	Comp. graphics		
3	Comp. graphics	20	HiFi technology		
5	Comp. hacking				
8	Comp. games				
10	Electronics				
13	Comp. music				
19	HiFi technology				
Practical crafts					
16	Photography	4	DIY	10	Cooking
19	Cooking	9	Photography	10	DIY
22	DIY	15	Cooking	19	Car maintenance
		20	Gardening		
Games					
15	Adventure games	–		–	
22	Math/logic puzzles				
24	Arcade games				
The arts					
3	Science fiction	1	Music, listening	2	Music, listening
6	Non-fiction	4	Reading, general	4	Reading, general
7	Music, listening	6	Non-fiction	5	Fiction
8	Reading, general	12	Fiction	6	Non-fiction
11	Fiction	14	Science fiction	15	Cinema-going
19	Cinema-going	19	Cinema-going	17	Theatre-going
Other activities					
12	Holidays	1	Holidays	1	Holidays
16	TV watching	3	Travelling	9	Camping
25	Astronomy	20	Pets, keeping	14	Travelling
		25	TV watching	17	Pets, keeping
				25	TV watching

whether structured or casual, and were more likely to have become involved in nurturing and socially responsible activities than the Dependents. In general they were more interested in all of the arts activities than either of the computer-owning groups, and were more interested in reading fiction and poetry than the others. They were also significantly more involved with all types of physical activities than the Dependents, especially those which were vigorous, outdoor sports.

Their interest in car maintenance, all manner of crafts, gardening and Do-It-Yourself indicated that they were extremely practical people who showed technological aptitude. Perhaps because of their practical nature and their apparent need for practical applications for their hobbies they may have been inhibited from purchasing a home computer, in spite of the fact that the majority of them had used one at work or while studying.

In general one might assume that the activities of interest to the Non-owners were more representative of those of the general population than were those from either of the other two groups. They appeared to be moderately interested in many different spheres, with no activities holding complete sway of the results. Nevertheless their reported activities cannot be assumed to be replicated by the whole population because of the predominance of well-educated males within this group. Within this group there were no doubt socially-phobic, technologically-biased, or physically-inactive individuals, however the features of the Non-owners' hobbies as a whole may be summarized as follows:

- Social
- Vigorously and competitively physical
- Practical and down-to-earth
- Artistic and cultural.

Hobby Profile of the Owners

The activities of interest to the second control group, the Owners, showed in some instances similarity with the Non-owners and in others they shared the same hobbies as the Dependents, indicating that they possessed certain characteristics in common with both (as had been anticipated).

Like the Dependents, they exhibited a far greater interest in puzzles, games and technological hobbies than did the Non-owners; although their interest levels were not always as high as those of the Dependents. However, they were also people who appeared to desire practical applications for many of their leisure pursuits, and held equal interest with Non-owners in the domestic activities of Do-It-Yourself, gardening and handicrafts. Although interested in the intricacy of high technology they obviously did not confine themselves to this area and were more interested than either of the other groups in the lower technologies of woodwork and metalwork. Although their technological bias could have been a precursor to their interest in computers, if they purchased one primarily as a tool and were frustrated by its lack of practical applications this could account for their subsquent waning interest in computing. Mere ownership of a computer appeared insufficient to sustain or increase their interest to the level of the Dependents.

Like the Non-owners, the Owners were far more interested in all manner of social and nurturing activities than the Dependents, and also in the sporting activities. Their interest in sports did not extend to the very vigorous activities of athletics, running and keep fit, of interest to the Non-owners, but they were not averse to competitive team sports and showed equal interest as the Non-owners in rugby, cricket and football. Although very interested in technological subjects the Owners did not confine themselves to such hobbies; they were social and physical beings who also appeared to show a practical bent in the activities they enjoyed. Their interests, showing characteristics both of the Non-owners and the Dependents, may be summarized as follows:

- Social
- Competitively physical
- Practical
- Technological and scientific.

Hobby profile of the Dependents

INTENSITY OF INTEREST IN HOBBIES

The results from the Leisure Activities Inventory and the interviews carried out with 45 of the Dependent group were able to confirm that the Dependents, although attempting similar numbers of activities as the control groups, were far more likely to reject many of them as uninteresting and to pursue the hobbies which were of interest to a much deeper level. One man described himself as a 'serial monomaniac', a person who would spend many months or years studying one particular area before moving on to another hobby which was perhaps closely related to it. The Dependents appeared to be specialists and far less eclectic in their interests than the control groups. They took most of their past hobbies very seriously, with computing merely being the last and best in a series of related activities. This was illustrated by the following quotes—the first from the wife of one of the interviewees.

> 'He's had many obsessive hobbies. He couldn't keep his hands out of the insides of cars—anything to do with wires, lens, valves—electronics, cameras, etc. and squash, bridge and wine-making. Everything is done to excess, he has to be an expert.'

> 'I've been hooked on things before—it's just a search for knowledge, to find out something—I would worry about not knowing. It's a whole new

dimension in function and capabilities in a machine, way beyond cameras, cars and motorbikes.'

'I've always been very heavily involved for a hobbyist—I've been into Go-Karts, radio astronomy, electronics, plane restoration, and I was a film buff. I prefer to do something thoroughly, until you are reasonably competent in an area you can't actually achieve anything.'

'As soon as I could read it became an obsession. I never did my homework and I didn't like stopping for meals. I studied languages from the age of 14, on my own, and was really obsessed by astronomy from 15—I read everything in the library. I taught myself about maths, the teachers were hopeless, and I'm still fascinated by classical Greek mythology and folk tales. I don't have time now—they don't hold my interest in the same way as computing.'

Hobbies of least interest to the Dependents

The most obvious and significant difference between the Dependents and the two control groups was in the number of activities which had been attempted less often or were of lesser interest to them. The majority of the activities which fell into this category were those with physical or social features.

The Dependents as a group showed almost total disinterest in sporting activities of any type. Most of the Dependents who had been interviewed gave the impression that they seemed to find sporting activities faintly ridiculous and totally pointless, and focused most of their antipathy or hostility upon competitive ball games. They preferred to be totally responsible for their own actions and not to be part of a cooperative team effort, citing any preferred physical activities to be ones where they could be self-reliant. However, the results showed that even for the non-competitive, non-cooperative sporting activities the Dependents' attempts and interest levels never exceeded those of the control groups, and even for the passive activities of spectating the Dependents were under-represented. This lack of interest in sports was evident even at an early age, as the Dependents had also reported themselves to be non-sporty while at school, preferring to spend their time on more academic pursuits.

Their lower interest than the controls in social activities was also confirmed, as had been suggested by the reported number of evenings spent socializing by each group. As with sports, the interviewees revealed that their attitudes towards social activities had also been formed early in life. The majority of the Dependent interviewees considered themselves to have been shy from an early age, and because of their early experiences would often take pains to avoid social situations which they would find stressful. They did not enjoy general get-togethers which they felt served little purpose, and such unstructured events as going to pubs, dances and parties were particularly shunned.

Entertaining within the confines of one's home and dining-out were two of the few forms of enjoyable socialization mentioned by the interviewees, who even then preferred them to be limited to small, intimate groups of special friends. Gatherings of like-minded people who were pursuing similar interests to their own gave the greatest pleasure—occasions where ideas not intimacies were exchanged. However, even this form of socializing was rare for many and was usually initiated by someone else, invariably the spouse. None of the activities which could be described as nurturing and socially responsible, nor even holidays and travelling, all of which usually have a high social content, were of particular interest to the Dependents.

Hobbies of greatest interest to the Dependents

The results from the interviews, substantiated by the reponses to the Leisure Activities Inventory, gave the overall impression that the Dependents were in almost constant need of positive, intellectual stimulation in their lives, not obtained through social or sporting activities. Even if they were able to achieve this from their daily work or studies, as many were, this need was rarely fully satisfied and their leisure activities were also selected to provide this type of stimulation. This was realized through personal study, usually undertaken solitarily, and the Dependents often devoted as much time and effort to their leisure pursuits as to their professions.

Of the three groups, the Dependents showed the greatest interest in activities which could be classed as intellectual games and puzzles. This enjoyment of puzzles was reflected during the interviews of many of the Dependents who expressed sorrow that they had not been asked to complete batteries of 'Intelligence Tests' for this research. Many stated that they loved challenging themselves in this manner, especially against time limits, and that such exercises were seen as pure pleasure. Quite a number were members of *Mensa* and therefore had had experience of completing two full tests under time pressure. Through past experience they were aware that their analytical, logical, intellectual abilities were those measured by such tests. Many puzzles and games demand considerable concentration, strategy and planning, and computer adventure games, which involve logical and 'lateral' thinking, were of especial fascination to them. Although not physically competitive they relished intellectual contests which were usually pitted against their own past achievements or against the devisors of the games. The Dependents' fascination with such activities was consistent with the pleasure gained from computer programming, hacking, breaking software codes and the refinement of commercial software.

Lack of interest in practical applications

The Dependents were shown to have attempted many of the various crafts listed in the Inventory. They, like the Owners, had been interested in model-making, but the interviewees revealed that it was the process of making or refining them rather than the finished models that held the Dependents' fascination. Similary, although some of the interviewees were very interested in cars, they were more interested in building or merely tinkering with them than in their maintenance and use. Unexpectedly, the Dependents' interest in photography was not found to be as strong as had been suggested by the responses to the General Questionnaire, and their interest levels were lower than those of the control groups. However, as well as being a technological activity, photography can be seen to contain very significant social aspects (see Beloff, 1985) together with physical and artistic characteristics, which were not of especial interest to the Dependents.

The Dependents' interests were therefore more cerebral than practical. Although dextrous and knowledgeable they were less interested than the control groups in the truly practical crafts demanding lower technological and more mechanical skills. Few of the interviewed Dependents saw themselves as 'handymen' and most stated that their preferred activities rarely had any recognizable useful outcome other than that of enjoyment and intellectual stimulation. Activities such as Do-It-Yourself and gardening were simply viewed as chores, to be avoided if possible in favour of the more esoteric activities.

Precursors to the Dependents' interest in computing

As had been expected, the Dependents did show greater interest than the controls in the computing activities, and the interviews revealed that for the majority this interest was founded long before they had direct hands-on experience. Two seemingly contradictory factors appeared to have been prerequisite for this intense interest in computers to develop. Firstly, many had had long experience with other forms of communication equipment and therefore had a practical, working knowledge of related devices, and secondly, from their interest in science fiction, they appeared to invest the computer with almost divine, often spiritual qualities. The computer was comprehensible and transparent but was also an omnipotent instrument which could enhance the abilities of man. Such factors lifted the computer from being a mere tool, used for processing the family budget, to something which was in itself intrinsically stimulating and exciting.

Prior experience in electronics, various forms of communication equipment and other 'high-tech' gadgets were significantly linked with the Dependents' interest in computing. The interviewees loved their detail and intricacy, their clean image and their power; a power which they could manipulate and control. For many this interest developed early in life. Although most were unsure whether their parents had encouraged these hobbies, they had been given the freedom to build, dismantle and explore all manner of electrical and electronic equipment at an early age.

The equipment varied from crystal sets to sound systems and programmable calculators, and two had built their own computers from kits long before microcomputers were freely available.

Although much of their interest centred upon communication equipment, the purpose of this communication was more likely to be for the expansion and interchange of knowledge than for any extrinsic practical purpose or for the ability to get to know other people. Stereo equipment was adapted and tuned to sophisticated levels, but the impression was given (often by spouses) that the quality of the sound was more important than the content of the music. Even the person-to-person exchange of amateur radio was exploited for the cataloguing and categorization of countries and cities with whom contact had been made. Similar factors often applied to communications over the computer networks. The need was for intellectual stimulation and the exchange of ideas, not socialization.

The second aspect which seemed to be a prerequisite for the formation of computer dependency for most people was the quality of fantasy and power with which the computer had been endowed. Once again this seemed to have developed long before they had used computers and appeared closely linked to their interests in science fiction and science fantasy. This was the favourite reading matter of the majority of both male and female Dependents, and had been raised to the level of philosophy for many. Although realistic about the limitations of modern computers, most were enthusiastic that information technology would develop to the stage where the abilities of such computers as '*Hal*', in Kubrik's film *2001, A Space Odyssey*, would be superceded. (Both Frude, 1983 and Turkle, 1984, among others had also observed computer dependents' great interest in science fiction.)

There was little doubt that the interviewees saw themselves as philosophers, lovers of wisdom and knowledge, and many appeared to devote their lives to the pursuit of this end. Their interests in the philosophies of others ranged from the classical and William Blake's *The Book of Urizen* (1979) to *Zen and the Art of Motorcycle Maintenance* (Pirsig, 1974). However, the most beloved tome of many was Hofstadter's *Godel, Escher, Bach: an Eternal Golden Braid* (1979), a treatise on the association between Bach's fugues, Escher's drawings, Godel's Theorem, and Artificial Intelligence. The combination of their interests in science fiction and philosophy, together with mythology and astronomy, gave the impression that the Dependents were visionary futurists for whom few reflections could be thought of as too idealistic or romantic. These attributes distinguished them from the control groups whose leisure activities appeared more down-to-earth and practical.

The artistic interests of the Dependents

Although predominantly scientifically oriented, the results indicated that the Dependents also appeared to show some artistic tendencies. On five occasions the interests of the Dependent matched those of the Non-owners, not only in philosophy but also in meditation, amateur dramatics, computer music and in writing stories and articles. The interview data were able to explain these interests more fully.

Meditation was seen by some of the Dependents as being a method for gaining spiritual peace in a social world with which they were not entirely comfortable and was linked to their earlier interest in other philosophies. Amateur dramatics, perhaps a somewhat surprising activity for the apparently less out-going Dependents, was seen by some as a means of fully expressing themselves within the confines of a safe and secure role. This activity shows a similar need for a fantasy life and to the Dependents' need for structured social activities which have a definitive purpose. In spite of the initial results which demonstrated that music was of great importance to the Dependents, these results showed that there was no difference in the interest levels between the groups, except when the music was electronically generated. The writing undertaken by the Dependents was discovered not to be fictional, as could be expected from some of the Non-owners, but invariably took the form of writing articles about computing, many of which had been published in the popular computing magazines and journals. Such results were therefore compatible with the overall profile of the Dependents' preferred leisure activities.

From these results, the Dependents can be seen as people who actively seek out hobbies which act as challenging puzzles to intellectually inquiring minds. Although rarely reporting themselves to be practical in outlook, the Dependents were skilful technologists who desired to explore and understand the physical world. Their interests centred upon the inanimate, upon artefacts over which they could exercise power and control. They expressed disinterest in the psychological, emotional and cognitive aspects of humans, including themselves, except where this impinged upon the development of artificial intelligence. They preferred the rational, predictable and logical aspects of the objects of their interest to the contrary attributes of the human race, and preferred to devote themselves to solitary pursuits which could fulfil their needs. The attributes of the Dependents' leisure activities may therefore be summarized as follows:

- Intellectual and exploratory
- Technological and scientific
- Fantastic and visionary
- Philosophical
- Non-physical
- Non-practical
- Non-social

For the Dependents, computing was seen as an activity sought out for its own sake and for the sake of intellectual stimulation. Many of their past hobbies had also been pursued with much the same fervour and enthusiasm, although invariably these had paled into insignificance since the advent of the computer into their lives. Their past hobbies were often studied for many years and had not been abandoned lightly, but their interest in computer had superceded these activities and seemed destined to remain of primary interest.

None of the 75 Dependent respondents reported that they had been dependent upon computers for less than two years and they gave a mean time of over five years (see Table 7.14). They also considered their dependency to be strong as well as abiding. When asked to make an assessment of their dependency in terms of a percentage commitment the results yielded a mean of over 80 per cent, and when 22 spouses assessed their partners' dependency, this percentage rose still further. Although purely subjective, these assessments of rates of commitment to computers did confirm the seriousness of their interest.

Table 7.14 Length of time and strength of dependency upon computers

Dependency upon Computers	Mean	S.D.	Range
Length of time dependent	5·6 years	4·5	2–23 years
Strength of dependency (self)	81·6%	11·9	50–100%
Strength of dependency (partners)	87·3%	13·9	50–100%

The data obtained from the investigations into the Dependents' leisure activities suggested that computing and other past hobbies were able to fulfil many of their psychological needs; needs which were assumed to differ from those of the control groups. Because of the manner in which the Dependents used their computers, they provided them with unlimited powers for ingenuity and exploration; if one machine was fully understood and mastered a bigger, more powerful, microcomputer was purchased. The ability constantly to stimulate and provide novel experiences seemed unique to the computer.

Computing, hobbies and personality

This investigation was able to confirm that the Dependents' intense interest in computers and computing had been influenced by their history of past activities and had not been made in a vacuum, and that the pleasure experienced when computing was well-matched with their needs. That hobbies and free-choice activities play an important role in the development of satisfactory and satisfying lifestyles has been well documented, with numerous theories as to the needs gratified.

The need for leisure pursuits and play

Auden (1948) saw pure acts of choice as necessary for the health of the ego, as retaliation against the instinctive desires which control much of our lives. That some of the Dependents held their basic needs in scant regard was reflected by the lack of importance they placed upon the basic needs for sleep, food and other creature comforts. Similarly, Bruner *et al.* (1976) suggested that play and leisure remove us from the serious aims of life. Bishop and Witt (1970) believed that hobbies provide us with attainable goals, a release of tension, and a means of relaxation and restoration, while Hebb (1955) saw play as a means of creating arousal by avoiding the states of monotony and boredom; all theories which were pertinent to the Dependents.

Play has also been shown to give the opportunity to experiment with behaviour patterns and to explore problems in a safe and stress-free environment (Sylva *et al.*, 1976). Puzzles of all types enable us to test our intelligence and tenacity (Huizinga, 1976), and this is well reflected in computing which has rules and protocols which have to be adhered to, but which can be pushed to the limit to test one's resources and prowess. Rules, although giving a game importance, cannot mask the fact that the aim of any game is inherently frivolous (Auden, 1948), and has no intrinsic rewards (Bruner *et al.*, 1976). It serves no other purpose apart from that of the act itself; well reflected by the Dependents' lack of practical end-products from their computing.

The Dependents' computing activities were inherently playful. Whatever the need for play, and this will vary from person to person and day to day, most authors attribute it with similar characteristics. Play is pleasurable (Bruner *et al.*, 1976), play is voluntary (Sylva *et al.*, 1976), play is not 'real' but steps out of real life into fantasy (Haun, 1965; Huizinga, 1976) and is satisfying in itself. If the activity does not satisfy these criteria it ceases to be play, and ceases to exist when it has any other purpose but that of enjoyment (Haun, 1965). That intense pleasure and excitement was experienced by the Dependents was reflected by their use of such expressions as 'getting a high' or 'a buzz' (presumably an adrenal reaction) when describing the feelings of pure elation and excitement stimulated while computing.

To many people, using a computer is work not play, and in many instances is tiresome, tedious and frustrating. These were not the attributes given to computing by the Dependents. Not only did many work with computers during the day but they also chose to spend their free time pursuing similar activities. Lucky is the person whose paid employment is seen to exhibit the same attributes as their play. Although some people enjoy leisure activities far removed from their everyday labours in order to satisfy their needs, others, because of positive reinforcements found in their paid employment, prefer to choose tasks very similar in nature to their work, and vice versa (Hagedorn and Labovitz, 1968).

Thus not all play is the same. Although some can be lighthearted and frivolous, Haun (1965) recognized that play, as well as stepping out of the confines of real life into the realms of fantasy, can also be as serious and as intellectually demanding as any vocation. Some leisure pursuits, such as those traditionally enjoyed by the Dependents, can involve fully as much study, determination and self-discipline, and perhaps more in the way of ingenuity, inventiveness and imagination, than many jobs. Therefore the games played will vary according to the types of autonomy and pleasure desired and with the needs and interests of the players. This therefore does not exclude the fact that what is pleasure for one can be purgatory for another.

> 'They (games) are only important to those who have the physical and mental gifts to play them, and that is a matter of chance. Granted that a game is innocent, the test of whether one should play it or not is simply whether one enjoys playing it or not, because the better one plays the more one enjoys it.'
>
> Auden (1948)

While the Dependents' hobbies were for fun and intellectual stimulation, the chosen hobbies of the control groups indicated that they gained pleasure and interest from activities which were inherently practical and useful, and less of their leisure time seemed to be devoted to fanciful play. Perhaps the control groups were more likely to have been driven by a sense of duty; duty which was obviously enjoyed and accepted as being an integral aspect of living in a social community. As Auden implied, a sense of duty prevents the activity from being play, but it does not necessarily prevent it from being enjoyable. Many of the Dependents felt little sense of duty to others, not even to their own family members. Although many felt guilty about the domestic chores now left undone, the family outings never taken or the lack of time spent with spouse or children, computing had become the priority.

The Dependents were playful, often childishly so, although this is by no means meant in a derogatory way. They still seemed to have conserved that wonderful freedom which allowed them to behave with abandon, perhaps because they had not been socialized to behave as 'adults' by the influences of others. Forty-year-old men still enjoyed 'zapping spaceships', and doing 'naughty' things like hacking into protected computer systems brought endless glee and delight. The following quotations sum up the Dependents' varied needs, satisfied by the computer:

> 'Initially it was a challenge—I didn't think I would ever manage it, then I discovered I could do it and master it. The sense of achievement—even if it's just an example in a book or to fix a small bug—it's marvellous. I always accept a challenge—I'm stubborn and determined and patient. The fact of typing a line of things in and seeing this magic thing do whatever you wanted it to—it's so fascinating—the instant feedback if it works.'

> 'It's an excellent companion when you're alone and it's of use. It gives me power. It's a machine that I can run—I can cleverly control it—it's your servant. It gives a psychological buzz—I lose track of time. It's a way of relaxing, like playing chess or bridge—both are also addictive.'

Because of these descriptions it was of little surprise that the interviewees had no desire to rid themselves of their dependency upon computers, and most actively wished they had more time in which to carry out their desired activities which gave them so much fulfilment.

The association between personality and hobbies

It was possible to deduce from the interviews that the computer was not only a toy to the Dependents but also a competitive rival to be dominated. This opponent, however, played the game by rules which were compatible with the cognitive style of the Dependents. It was seen as rational, logical, reasonable and intelligent, and through perseverance, prowess and the use of protocols, the Dependents were able to dominate the battles. Such need for control over the inanimate, found by McClelland (1962) to be a factor of the scientific personality, was well reflected by some of the Dependents who used it as a substitute for their lack of control in other areas of life. The Dependents frequently made comparisons between their interactions with the computer and with other people, and invariably the latter were found wanting and seen as far more difficult to cope with or comprehend. Investigations by other researchers into the area of leisure pursuits were able to show that the Dependents were not unique in their needs, nor in their preference for an interaction with a machine rather than other people. Research has shown that the types of leisure pursuits enjoyed have been found to differ between the sexes and according to interests, aptitudes, personality types and intellectual abilities. The Dependents' choice or avoidance of certain activities were found to be consistent with other research findings, especially those which highlighted differences between those with a scientific/technological bias and others.

In her extensive studies of academic types, Roe (1953) found that scientists (physicists and biologists) differed greatly from psychologists and anthropologists on many variables. The scientists experienced far less personal interaction early in life, were more shy, less rebellious and had often been isolates as children. They found personal relationships difficult and on the whole uninteresting, leading them to choose recreational activities which were normally solitary.

She believed that one's hobbies tend either to be person-centred or object-centred depending upon the personality of the individual, with the Dependent respondents therefore falling into the latter category. Later research (Roe, 1957) indicated that an orientation of interests which was not person-centred early in life could lead on to the development of scientific and technological biases later. She did not believe this interest

join clubs like the Chess and Wargames or Model Railways clubs. [Computer Studies] students are very plain and dull in appearance and are the sort of people who wouldn't stand out in a crowd.'

Perhaps surprisingly, these two quotations were from computer science undergraduates describing themselves and their fellows in the computer studies department. Physical education and English students who also took part in this study all tended to give similar descriptions of them. Computing students were seen as very intelligent, quiet, serious and withdrawn, with hobbies appropriate to their personalities.

The Dependents' hobbies were also found to be similar to those identified by Neulinger and Raps (1972) for a group of American *Mensa* members. When comparisons were made between them and an unmatched sample, the *Mensa* members preferred activities which involved analysis and reflection while the control group preferred nurturing, social and affiliative activities. The *Mensa* members tended not to see either work or leisure as social events and sought out 'self-centred activities' wherever possible, as did the Dependents.

Although these studies appear consistent, other investigations which have relied more heavily upon empirical, statistical analysis have in the main been unsuccessful in their attempts to correlate leisure activities with other factors. Thorndike (1935), although one of the first in this field, can be seen to speak for those who have followed. He felt that the differences in interests must lie in the 'genes of men', but concluded that it was 'not easy to discover what they are'. Although a lot of work has centred upon vocational interests, little has been achieved in the area of leisure and recreational pursuits, and that which exists appears to concentrate upon isolating factors rather than producing theory.

A number of investigations have shown that mechanical, scientific and athletic pursuits are of greater interest to males while females prefer social and aesthetic activities, (see Berdie, 1944; Hammond, 1945; Sandall, 1960; Nias, 1975); and Strong (1943) believed that people's interests became less physical and more aethestic and cultural as they grew older. Four studies have investigated the correlation between the interests of parents and children (Forster, 1933; Gjerde, 1949; Strong, 1943; and Nias, 1975). Their correlations were extremely low, often below 0·2, and few had been able to correlate leisure activities with psychological and social factors. Gjerde (1949) found a correlation of 0·31 between boys and their fathers for scientific interests, and Nias (1975) concluded that fathers low on psychoticism and introversion preferred academic interests and would also have introverted sons. However he could only find 'correlations typically around 0·15' for the numerous factors he examined.

Furthering the work of Tinsley *et al.* (1977), which assumed that people seek out situations which satisfy their personal needs, Furnham (1981) investigated whether the personality variables of neuroticism and extroversion were correlated with the choice and avoidance of leisure activities. It is perhaps of no surprise that he discovered that extroverts took part in significantly more social and physical activities than did introverts; he also discovered that introverted neurotics enjoyed reading and avoided competitive games and social activities. Furthermore, when choice and avoidance of particular activities were examined he concluded that avoidance patterns were determined by personality factors which reflected a person's inadequacies rather than by their ability to perform the activity.

The results from these studies were various, with some obvious, confirming common opinion, and others more enlightening. From them, and from the profile of hobbies exhibited by the Dependents, it was consistent that the majority of them were male, intelligent, enjoyed 'masculine' activities, and that some used their hobbies as an escape from social relationships. The results from the Leisure Activities Inventory were able to show that the hobby profile of the Dependents did differ from those of both matched control groups, and that in addition they were more likely to practise fewer hobbies to a greater depth and with greater interest. The Dependents' enjoyment of computing had not developed in isolation but bore a strong relationship to their past hobbies and shared many of the same characteristics.

Displaying such hobbies, the Dependents were expected to show convergent modes of thinking, developed as an effective, safe and successful means for dealing with their environment. Believing this mode of thinking to have developed early in life it should, if conforming to theory, be as a reaction against cold, neglecting, rejecting or disinterested parents whose approval was shown for ability rather than in an open, loving way. In addition, their avoidance of the social and physical activities was not only thought to reflect inadequacy in these areas, but also to indicate the presence of the trait of introversion.

The following chapter describes the research carried out in order to discover whether any of the above suppositions held true. Through the media of interviews, questionnaires and psychological tests, aspects of the Dependents' family backgrounds, attitudes and personalities were investigated to discover what might have led to their psychological dependency upon computers.

To conclude, two differing attitudes, both of which might be seen to describe the leisure activities of the Dependents, are posed.

> 'To many who are contemplative there is nothing more worthwhile in life than the increase in knowledge for its own sake.'
>
> Baker (1939)

> 'For a scientist of today to confine himself to purely individual work and not to attempt to integrate his work with that of his fellows is deliberately to restrict his capacity for discovery and to treat the overcoming of unnecessary difficulty as something to be pursued for its own sake.'
>
> Bernal (1939)

Chapter 8
Social and psychological influences: What might have led to the development of computer dependency

> 'He moves away from people . . . He feels he has not much in common with them, they do not understand him anyhow. He builds up a world of his own—with nature, with his dolls, his books, his dreams'
>
> Horney (1946)

Introduction

The hobby profiles of the Dependents suggested that their leisure activities had traditionally been object- rather than person-centred and seemed able to satisfy their need for control within their lives, a need perhaps frustrated under other conditions. The literature suggested that this bias of interest develops early in life, within the home environment, which together with specific personality traits and cognitive styles provides an effective method of dealing with life. Quite a number of the interviewees confirmed that their social and family relationships were problematic, and that they somehow considered themselves to be different from other people and outside the mainstream of life.

This chapter details the work undertaken to determine whether and which social and psychological variables had been influential in the Dependents' later dependency upon computers. Various issues were explored; attitudes to computers and people were formally investigated, as were family relationships, social experiences while young and when adult, and personality traits. The results and ensuing discussion have been divided into sections under the following headings:

- Attitudes towards computers and people
- Family life
- Social experiences
- Personality factors.

Attitudes towards computers and people

During the course of the interviews some of the respondents made direct comparisons between people and computers. Computers were being ascribed human characteristics, even by those who did not consider that they anthropomorphised the computer; conversely people were described in terms which were often diametrically opposed to those used to describe computers. There was an overall impression given that people were 'difficult' to interpret and understand, while computers were 'easy'; not an attitude universally held one would imagine. There was therefore a need to discover if this was the overriding view of the whole group of Dependents and to determine if this differed from the opinions of the two matched control groups. Measures were devised in order to determine whether the three groups differed in their attitudes towards computers, and in their attitudes towards people.

Rating scales, when used to examine attitudes, enable one to measure with some degree of reliability differences in the degree of attitudes held. From the data obtained from the interviews it was possible, using a simple content analysis, to extract the adjectives which had been most commonly used by the Dependents when computers and people had been described. Following the semantic differential technique developed by Osgood *et al.* (1957), sets of bi-polar adjectives were isolated and five-point attitude scales were designed.

Two scales were used and both were identical but for their headings; the first dealt with attitudes towards computers, the second attitudes towards people. To avoid response biases the adjectives were arranged alphabetically with a mix of both 'positive' and 'negative' qualities from one side of the scale to the other. Pilot studies were carried out to test for comprehension of the poles, with alterations and reductions in the number of adjectives made where necessary. The scales were sent to all three sets of respondents, and the analyses were carried out upon the results from the 75 Dependents who had completed all the postal measures and upon those of the two groups of controls matched with them. Each scale, and each of the bi-polar adjectives, was analyzed individually, with all analyses being carried out upon the frequencies of responses for each of the five points of the scale between the three groups.

Attitudes to computers

Of the 28 bi-polar adjectives on the *Attitudes to Computers Scale* there were 22 significant differences between the groups, with 14 of them showing significant differences between all three groups.

The results yielded a definite trend, showing the attitudes of the Dependents towards computers to be more positive than those of the Owners, which were in turn more positive than those of the Non-owners.

In only one instance were the results from the two groups of computer owners, the Dependents and the Owners, found to be similar. In seven instances, the two control groups held attitudes towards computers similar to each other, although different from those of the Dependents.

In order to illustrate the differences between these significant results effectively, profiles of the group means are shown in Figure 8.1. For additional clarity, the scales have been reversed where necessary and have been grouped into three major sections according to where differences were found between the groups. The scales are ordered to show increasing values of the means as determined from the Dependents' results, with the 'positive' attributes listed on the left-hand side. Where no differences were observed between two groups their combined mean scores were determined and are illustrated as running in close parallel.

Attitudes to people

The results from the *Attitudes to People Scale* showed that in this instance there were no significant differences between the results from the two control groups for any of their reported attitudes. However, nine significant differences were observed between the Dependents and the two control groups. The significant results are illustrated in Figure 8.2; once again showing the mean scores of the Dependents in increasing order of value, and combining the group mean scores of the two control groups. The left-hand side of the figure contains the 'positive' attributes.

Discussion

The results from the *Attitudes to Computers Scale*, although showing a great number of significant differences between the groups, also indicated that all three groups tended to hold positive attitudes towards computers. This may be observed by the physical representation of the group means in Figure 8.1, where the majority of the results lay to the left hand side of the mid-line, the side which may be considered to contain the positive poles. The profiles are almost parallel to one another, indicating that the results from the three groups differed only by degree, and were not the antithesis of eath other. One may conclude from these data that the three matched groups, consisting in the main of well-educated, young adult males, all tended to credit computers with 'positive' characteristics.

More similarity between the results of the two computer-owning groups had been expected, but in only one instance were their results aligned. They described computers as easy while the Non-owners tended to find them difficult. Although the vast majority of the Non-owners had used computers, it was perhaps for this reason that they chose not to become computer owners in spite of the fact that they found computers

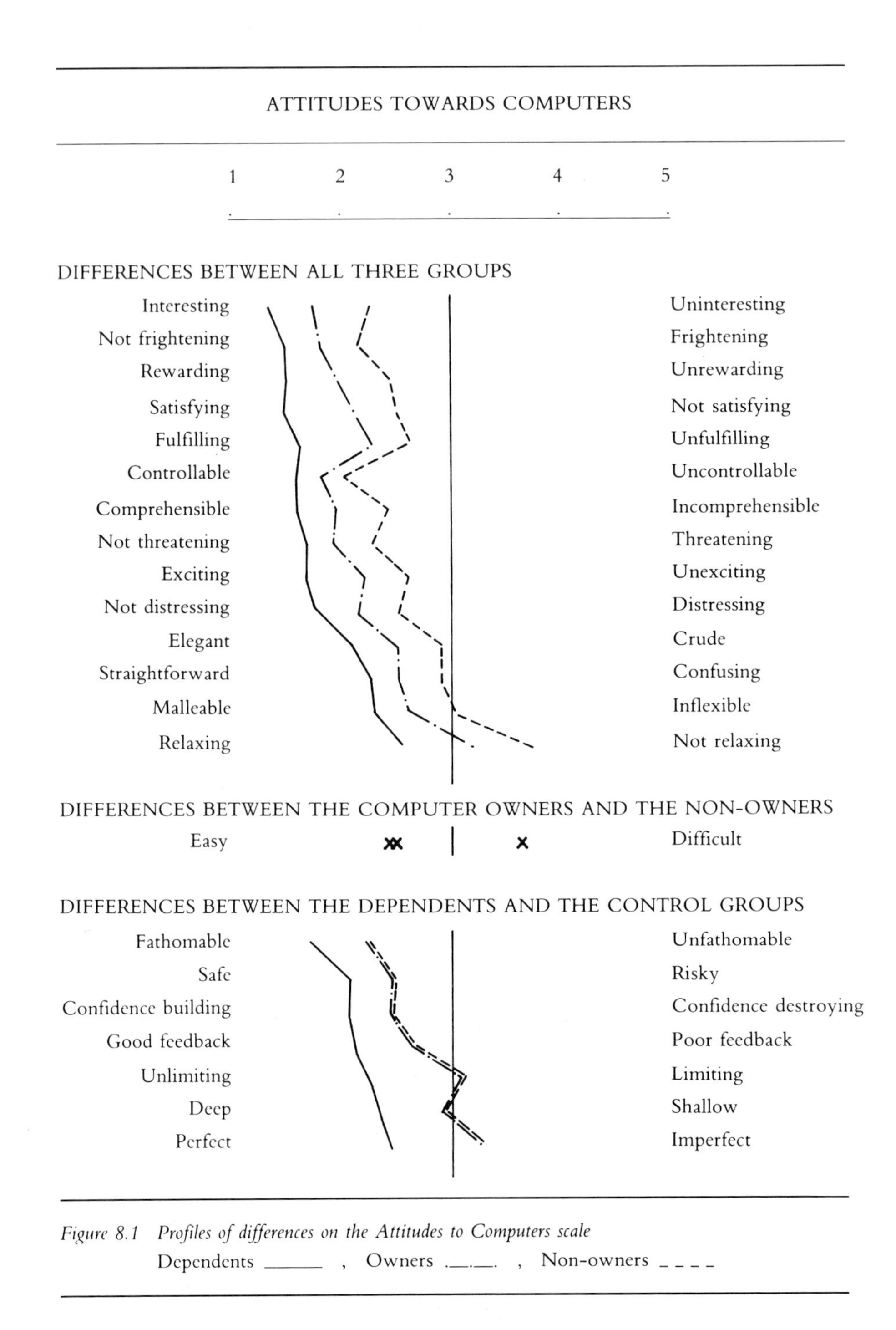

Figure 8.1 Profiles of differences on the Attitudes to Computers scale
Dependents ——— , Owners —·— , Non-owners - - - -

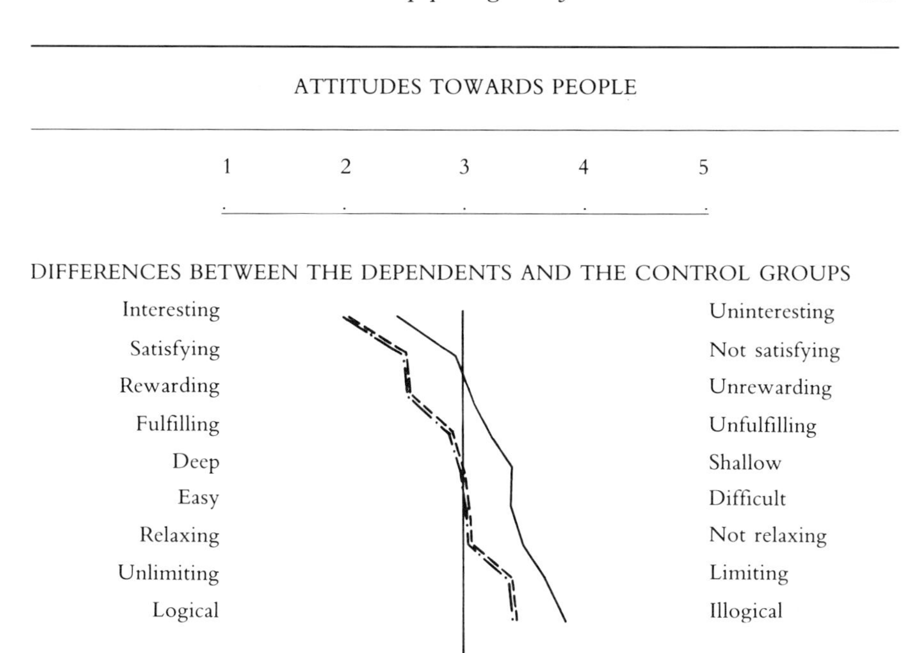

Figure 8.2 Profiles of differences on the Attitudes to People scale
Dependents ——— , Owners .—.—. , Non-owners - - - -

interesting and controllable. Where similarities in attitudes were found these were in the results of the Owners and the Non-owners. These corresponded with the results from the Leisure Activities Inventory, which usually showed the interest levels of the two control groups in computing to be equal to each other, and lower than those of the Dependents.

Unlike the Dependents, the two control groups did not find computers relaxing and this attitude would be likely to inhibit pleasurable use of any artefact, however interesting. There is little doubt that most naïve users experience problems when first interacting with a computer, and if these difficulties are viewed as threatening and frustrating, rather than challenging, skills and interest are unlikely to develop easily. Perhaps the control groups' descriptions of computers as limiting and imperfect suggested that they found them inadequate for their needs, since interpretations of their hobby profiles assumed them to be practically biased. Because most of the Dependents used the computer as a vehicle for exploration and self-education, uninhibited by the desire for practical end-products, it was more likely that computers would be viewed by them as less limiting and more perfectly suited to their needs.

One may conclude from these results that the Dependent group gained far more from their computing experiences than did the control groups.

Many of the descriptions given in the interviews were confirmed when all of the group were asked for their opinions. The Dependents' ratings of their attitudes to computers demonstrated that the qualities attributed to the computer, their interaction with it, and the feedback gained, were viewed as far more positive, pleasurable and unintimidating than were those of the control groups. The Dependents' interaction appeared to be seen as safe and secure, while for the control groups their responses seemed to express less confidence. Nevertheless, in most cases the control groups also looked upon computers favourably, but the high proportion of scales which gave significant differences between the groups confirmed that their interest levels were not as high as those of the Dependents.

Where differences did occur between the groups in the results from the *Attitudes to People Scale* all showed the attitudes of the Dependents to be significantly different from those of the two control groups, and in all instances their responses were more 'negatively' biased towards people. These results correlated with those from the Leisure Activities Inventory which showed the Dependents to have lower interest levels in social activities than the controls.

Figure 8.2 clearly indicates that the profiles from the groups again follow very similar trends to each other, but in this instance the majority of the results were less extreme and most lay about the mid-line of the scales. Thus all three groups saw people as somewhat illogical and limiting, but the control groups' results demonstrated that they also viewed people as deep, relaxing, fulfilling and rewarding, attitudes not shared by the Dependents who found them less interesting and far less satisfying. Their more negatively-biased opinions could be assumed to reflect accurately their analyses of the feedback they gained from their own relationships, which during the interviews had already been suggested to be less than satisfactory.

One may conclude from the results for these two scales that there were indeed distinct differences between the groups in the way they viewed computers and also in the way they viewed people, attitudes which are assumed to have developed through past experience. Confirming the information given during the interviews, the Dependents as a whole seemed to mistrust people and appeared to have less than satisfactory relationships with them. If this were so it is perhaps understandable that some deliberately chose to turn to a machine which had the ability to simulate an interaction seen by them as friendly, available, accepting and non-judgmental, while also providing the intellectual stimulation they so obviously needed. Perhaps this accords with the description by Fromm, (1941):

> 'In order to overcome his sense of inner emptiness and importance he chooses an object on to whom he projects all his human qualities: his love, intelligence, courage, etc. By submitting to this object, he feels in touch with his own qualities; he feels strong, wise, courageous, and secure.'

The interviewees seemed to prefer computers to people, and when comparisons were made between the 75 Dependents and the matched control groups these suppositions were substantiated. Further exploration investigated their early experiences within the family to determine if these had been influential in the formation of their attitudes. Had they been brought up in an environment which encouraged object-centred biases (Roe, 1957) or which rewarded deeds rather than the person (Getzels and Jackson, 1962)? Did they view their parents differently from other matched groups of people?

Family life

When asked to describe their family backgrounds and their parents, most of the Dependents did not consider that their upbringings were particularly out of the ordinary. They were usually described as typically middle-class and quiet, if sometimes rather isolated. Nevertheless, many did report trauma in their early lives, and although most described their fathers as cold and distant, as had the undergraduate students, it was not known whether this was in any way unusual. Blendis (1982) indicated, from in-depth interviews with men about their own fathers, that of 30 men from the ages of 30 to 50 years old, only three reported 'positive relationships without reservation', seven reported relationships which were 'quite warm, but with reservations', while the majority, 18, reported negative relationships with their fathers 'without warmth or closeness'. The reports of the Dependents may therefore not have been at all unusual, merely typical of the society in which we live.

The interview data were highly subjective, and being retrospective were therefore prone to memory lapse and distortion. Nevertheless, it was the Dependents' perceptions and feelings about their parents and their early life which were requested, and the investigation of family life was deliberately kept very broad in order to discover where differences lay between the groups.

Using the same techniques as had been employed for the previous scales, two identical scales were designed using the descriptions employed by the Dependents during the interviews. Each scale contained 22 bi-polar adjectives, one for the evaluation of each parent. Answers to certain multi-choice questions in the Final Questionnaire about families and backgrounds were also used. Differences were looked for between the Dependents and the two control groups, and data from the 45 interviews with the Dependents were used to support the comparative results obtained.

All analyses were carried out upon the raw data, although percentage values are used within the text. Significance levels are indicated within the tables by $\star\star\star = p < 0{\cdot}001$, $\star\star = p < 0{\cdot}01$, $\star = p < 0{\cdot}05$, and – = non-significant.

Results

None of the results from the questions pertaining to the family showed any significant differences between the three groups (Table 8.1). The majority of all three groups reported their mothers to be the most important relative in their lives, about one fifth reported their fathers, and a few their siblings and other relatives. The majority were closer to their mothers than their fathers, but most reported that they were not the favourite child of either parent with most stating that there were no obvious favourites within the family. The results were almost evenly divided between those who did identify with the parent of the same sex as themselves and those who did not, and the majority of each group stated that their parents were happily married. There were also no significant differences between the groups in the age at which they had left home, with the groups showing a mean of 19·8 years.

Table 8.1 Questions concerning family relationships

Questions of family life	Significance	Comments
Most important relative	—	Invariably mother
Parent closest to	—	Usually mother
Favourite of one parent	—	Majority were not favourites
Identify with same sex parent	—	Evenly divided
Parents happily married	—	Majority were
Age left home	—	Mean 19·8, S.D. 6·9

The analyses from the *Attitudes to Mother* and *Attitudes to Father Scales* showed differences for three variables on each of the scales, with all but one showing these differences to lie between the Dependents and the controls. These differences, illustrated by their mean results, are shown in Figure 8.3.

The mothers of the Owners were shown to be more outgoing than those of the other two groups, while the mothers of the Dependents were shown to be more dependent than those of the controls, and their fathers to be more authoritarian but also more encouraging. Most significantly, both the mothers and the fathers of the Dependents were shown to be more distant than those of the control groups.

The results from the interviews confirmed that the distance felt between the Dependents and their parents had had a very profound effect upon them when they were young, and often continued to cause them pain, anger or resentment. Many still felt estranged from their parents, and found it very difficult to talk about their early years during the interviews, with one man becoming particularly distressed. Typical descriptions of their fathers were reflected in the following quotations:

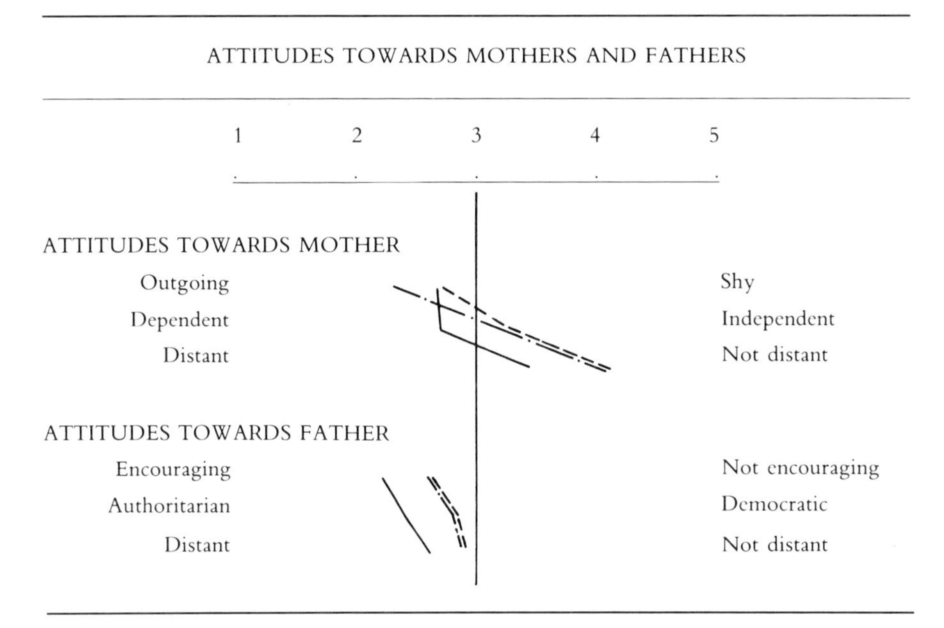

Figure 8.3 Profiles of differences on the Attitudes to Mother and Attitudes to Father scales
Dependents ______ , Owners .__.__. , Non-owners _ _ _ _

'He was strong willed—a patriarch—aloof. He was totally cold—estranged from me. He never carried me, or kissed me or talked to me. He would discipline very violently.'

'He's very much an academic. Doesn't like personal relationships very much. He was distant, especially to me.'

'Terribly strict, away from home a lot, an egotist. I couldn't stand him, he made me very repressed and timid.'

'He makes things, DIY, has lots of interests. He's not open, I can't talk to him. He's intolerant.'

'Obstinate, very much into himself, self-interested. Very much one to be concerned with his own interests and not yours. Distant, cold—stubborn, extremely so.'

The chief role of most fathers appeared to be as 'bread-winner' and disciplinarian (sometimes violently so) rather than carer, and they were often seen as peripheral members of their families. Although a few did describe their fathers as openly caring, this was not the overall consensus by any means. There were suggestions from some that their own fathers, like themselves, tended to be absorbed in their hobbies leading them to become estranged even when at home, and others were 'workaholics', who were rarely at home and often only saw their children at weekends.

There was far less agreement when the Dependents described their mothers. They varied from being seen as cold to over-protective, dominating to powerless, stable to mentally ill, but the suggestion that the love within the families was undemonstrative was often apparent.

> 'She was strong, stronger than father. Not tremendously caring, but I didn't suffer. We had very few visitors to our house, it was a quiet upbringing.'
>
> 'Very restrictive. She couldn't care less about me, but wanted me to do well. Our relationship got worse as I got older.'
>
> 'Very powerless. She suffered and put up with it. She was kind and loving, always on our side. She was a kind disciplinarian, but neurotic and depressive.'
>
> 'She was most things I'd expect of a mother—a nice mum. She lacked demonstrative affection but I think I was loved. She didn't let me grow up, and still does lots of cooking for me, for the freezer.'
>
> 'Very much a mother, very loving and dependable. She didn't encourage any of us to be independent, she needed to be a mother, needed to be adored.'
>
> 'That's difficult. She was small and dark—a manic-depressive. She was the best mother you could have when she was manic, we used to do such interesting things. I never cried when she died (she committed suicide when he was 11), I had no emotions from that age until I met my wife.'

Although none were asked for their sexual 'orientation', and similar numbers of the Dependents and the controls had married, one of the interviewees volunteered the fact that he was homosexual. His description of his relationship with his mother was abnormal; she, like one of the mothers previously mentioned, was also a manic-depressive.

> 'I didn't feel I had any identity apart from her. I was very attached. I felt I was part of her—I was part of her—I didn't realise I was a separate person until I was quite old. She said I belonged to her. I had no control over anything I did, I was totally under her influence and control. She was so suggestive—I would always agree—I could never think of anything better. She was very strict, she's a manic depressive, she drinks—she was very domineering, but she was really interesting . . . My father had nothing to do with her. He was very distant, almost non-existent. I only saw him on Sundays, I never really saw him . . . I went to bed with my mother—sexually, not really voluntarily. I'm confused about my mother—it may not have happened.'

He had broken off all contact with his mother at the age of 18; 'It was a clear-cut decision, no slow withdrawal, I had to break.' He had contemplated suicide many times and 'seriously attempted it' once.

Although it is not always possible to appreciate the depth of feeling that was engendered in the interviews by these quotations, it became very apparent that the influence of their parents remained very dominant within many of the Dependents' lives. Their general descriptions of their family backgrounds tended to be reflected by the comment, 'It was middle-class, fairly happy but isolated.' Few had homes which were visited by many friends, with the family unit being self-contained and therefore providing little opportunity for social interaction. Although the majority of the interviewees had siblings, they had tended from an early age to be reclusive in their interests and rarely seemed to share activities with their brothers and sisters. This may have been due to the confounding factor that most of the Dependents were first-borns. Their sibling relationships appeared to be unimportant rather than distressing or damaging.

The greatest concensus of opinion about the family backgrounds of those who had been interviewed seemed to centre upon the poor relationships which existed between themselves and their fathers.

The role of the family

Information about the Dependents' mothers and fathers gained during the interviews had been felt to be atypical of the general population. There seemed to be an uncommon proportion of interviewees reporting poor relationships with their fathers, with an additional incidence of people stating that their mothers were either not openly affectionate or conversely were overprotective, with a few of the mothers reported as suffering from some form of mental illness or instability.

Examination of the raw data from the two scales indicated that within each group there was wide variation in their ratings of their parents for each of the five-point attitude scales, and in all instances the neutral point of central tendency was only used by a minority of any group. This indicated that the respondents were not inhibited from using both the 'positive' and 'negative' ratings for their parents. Overall mothers were attributed with far more of the nurturing and emotional characteristics and less of the dominant, stable characteristics which were given to fathers. Although the overall results suggested that the respondents' relationships were poorer with their fathers than their mothers, one cannot conclude from these results that the majority of father/son relationships were as poor as those found by Blendis (1982). However, the results which showed the Dependents' ratings of their parents to be different from those of the control groups were seen to be highly relevant to their subsequent attitudes towards people in general, as well as being influential in the development of their object-centred biases.

The Dependents' fathers were seen as significantly more distant and authoritarian than those of the controls, but also as more encouraging. Although such encouragement could be seen to indicate a more positive and caring relationship between father and child than for the controls, this was not corroborated by the remaining significant results, the interview data or the literature. Encouragement is a term invariably used with reference to the performance of deeds and actions, and for this reason it can be seen to accord with the assertion of Roe (1957), Getzels and Jackson (1962) and Hudson (1966) that children are more likely to develop convergent, object-centred patterns of thinking if the task rather than the person is approved of and rewarded. The results from the interviews demonstrated that the performance of activities was the easiest or only method for the Dependents to gain positive attention and acceptance from their fathers.

That the Dependents' fathers were also described as more authoritarian and distant than those of the controls, appeared to bear out their descriptions of their fathers as cold, aloof and unavailable (except when contact was task or punishment-related). Such results suggest that the Dependents' fathers were less nurturing and caring than those of similarly matched groups, but their mothers were also seen as more distant. This fact could be assumed to have a very significant bearing upon the subsequent development of the child. In addition, although the mothers of the Dependents did not appear from the results to be more anxious, neurotic or frustrated than the mothers of the other two groups, they were described as more dependent, a factor which has been linked to distant fathering on a number of occasions.

Biller (1971) showed that where fathers were ineffectual or took little part in the nurturing tasks, maternal dependency was greatly increased leading to sons being rewarded for displaying mutually dependent behaviours, which were thought to interfere with 'their son's psychological development'. This in turn inhibited independence which was believed to hamper peer relationships and heterosexual development. The latter was seen to have occurred in at least one instance among the Dependents, and maternal over-protection was apparent for others. Although never specifically mentioned by other interviewees, the highly physical relationship which had developed between this mother and son was not found to be exceptional. Wylie and Delgado (1959) found many instances of mothers and sons sleeping in the same bed when the father was absent, often into adolescence, and Neubauer (1960) confirmed that such sexualised relationships could lead to severe problems with sex-role development of the child. That such relationships, which involve a high degree of father absence combined with a strong attachment to the mother, may lead to homosexuality is well documented, (Freud, 1947; Bieber, 1962; Bene, 1965). This case history was not typical of the other Dependents' descriptions of their parents, but in its extremity illustrates the severity of the problems which may occur.

Biller (1969), in other investigations into masculine development and role identification, observed that in families in which the mother had a more dominant parental role, the father was then likely to dominate his sons in a very restrictive manner. This also was shown to occur in quite a number of the Dependents' childhoods. Biller believed that fathers needed to be both masculine and nurturing for masculine development to be effective for sons. This nurturing element was found to be restricted if the son's appearance and obvious capabilities did not show high levels of intellectual and physical ability; the latter an unlikely quality for the majority of the Dependents, which could go some way to explain the difficulties between fathers and sons.

Positive father-son relationships, especially those in which the son identifies closely with the father, have been shown many times to have positive effects upon peer relationships and self-confidence (Rutherford and Mussen, 1968; Leiderman, 1959; and Cox, 1962). Conversely, poor relationships between father and son, leading to lack of a secure masculine orientation, were seen to prevent successful peer interactions (Gray, 1959), and the presence of a dominant mother was shown to increase such difficulties (Hoffman, 1961). Mitchell and Wilson (1967) showed father-absent boys more frequently to be rejected by their peers, and Jacobson and Ryder (1969) stated that the son would lack maturity and interpersonal competence, with marriages of their own being devoid of intimacy.

Based upon the administration of Cattell's 16PF questionnaire to groups of students, Siegelman (1970) found that males who had felt rejected by their fathers were timid, suspicious, insecure and tense, while those who had neglecting fathers were aloof, self-sufficient and tense. There were far fewer correlations between personality traits and parental interaction for females, but those with rejecting, neglecting fathers tended to be highly self-sufficient.

Deprivation in fathering, either through absence or low interaction, has similarly been seen to lead to fearfulness in social situations (Spelke *et al.*, 1973) to lack of trust in others (Santrock, 1970), depression (Malmquist, 1970) and to obsessive compulsive neurosis (Tseng-Wen-Sheng, 1973). Maturity and a sense of responsibility to others were also affected by deprivation of a nurturing father, and Bronfenbrenner (1961) found such children to be less likely to take any leadership roles or responsibilities when at school. Ineffectual fathering had been shown to lead to academic underachievement (Blanchard and Biller, 1971), and although this was not apparent within the Dependent group, some might not have reached their full potential.

During the interviews many of the interviewees stated that they *supposed* that their parents had loved them, but many were also very unsure whether this was really so because of the lack of outward manifestations of that love. The results from the scales did show that the Dependents described their parents as equally as loving as those of the controls, but if their love was non-demonstrative, reserved and restrained this could lead to the child subsequently learning to depend upon their own resources far more than would other children, and could lead to mistrust and fearfulness of human relationships. Distant, cool fathering has been shown to have profound effects upon children, and in this instance where both parents exhibit this characteristic, one could assume there was a far greater likelihood for this to lead to poor peer relationships (Gray, 1959; Biller, 1971), low self-assurance (Kotelchuck *et al.*, 1975) together with insecurity, tension and self-sufficiency (Siegelman, 1970). Such results can therefore be seen to explain, if only partially, why the Dependents appeared to have turned away from social interaction and relationships early in life, preferring to invest their trust and their energies in the inanimate.

Although the results showed the Dependents differing from the controls for these attitudes, a wide variation of responses was also present within the Dependent group, with a few yielding positive attitudes towards their parents. One could therefore assume, according to Roe (1957), that although the Dependents were object-centred rather than people-centred in their dealings with the world, not all came from cold and rejecting environments but that some were from the more casual families who tended to neglect their children. (Results discussed in Chapter 9 substantiate that there were different types of dependency.)

It had not been possible to interview the controls, nor was it possible to send standardized psychometric tests through the post to measure in greater detail how the controls differed from those of the Dependents. The measures used had remained very general, and although there were few differences found between the groups when their family backgrounds were examined in this fashion, I do however believe that the lack of outward warmth and affection, put forward so strongly by many of the interviewees, led to the Dependents being very unsure of their parents' love for them. This could be seen to form the basis of their subsequent aloofness and shyness, their mistrust of others and their apparent self-sufficiency and reliance upon inanimate artefacts. These factors appeared to be well confirmed when other social relationships and experiences throughout life were investigated, and when tests of personality were administered to the interviewees.

Social experiences

Further research was undertaken in order to try to determine whether there were observable differences between the social lives of the Dependents and the control groups. This investigated various aspects of their early social experiences and their attitudes towards school, as well as their adult social experiences and their attitudes towards work. If social problems did exist for the Dependents, and if they were more object-centred than the controls, it was felt this would be reflected in their attitudes towards activities other than their hobbies and in their dealings with other people throughout life.

Multi-choice questions included in the Final Questionnaire were used to determine whether the social experiences of the Dependents were typical of those showing similar demographic characteristics. The interview data were used to substantiate and elaborate upon by these results.

Schooling and early social experiences

The results from the questions which addressed various aspects of schooling and childhood are summarized in Table 8.2. They showed that the majority of all three groups found their academic work easy, felt themselves to be on a higher intellectual level than most of their peers, and had attended the same types of secondary schools. Most were educated at selective schools, whether state, private or direct-grant, where a few from each group had been boarders, while a small proportion had attended secondary modern, technical or comprehensive schools. Although the majority of all three groups enjoyed their secondary education, the Dependents were significantly less likely than the control groups to have enjoyed their primary school years. Few of any group reported feelings of rebelliousness when young, and no one group was more likely than another to have taken posts of responsibility at school.

Table 8.2 Schooling and early social experiences

Questions on schooling	Significance	Comments
Found academic work easy	–	Majority found work easy
Intellectual level with peers	–	Majority felt they were superior
Type of school	–	Varied
Enjoy primary school	***	Dependents less likely to enjoy
Enjoy sescondary school	–	Majority did enjoy secondary school
Rebellious as teenager	–	Majority were not
Take on school responsibilities	–	Approximately half had
More serious than peers	**	Dependents felt more serious
Felt different from others	**	Dependents felt different
Shy as a child	**	Dependents more likely to be shy

The remaining results showed consistent differences between the Dependents and the controls. The Dependents were significantly more likely to describe themselves as different from their peers and to be more serious and shy than the control groups.

The interviews with the Dependents had given the impression that for many their school years were very unhappy. Although usually academically successful, socially many felt extremely isolated from other children because of their inability to establish good social relationships. Most reported themselves to have been very shy from an early age and some to being ostracized by their peers. They did not share the same hobbies, were not the types to become 'gang' members, and felt that their superior academic abilities were considered unacceptable to the other children. Some even described themselves as feeling 'alien' and very different from their peers.

The situation tended to improve during their secondary school years when some found it easier to make friends. However, even here there was great variation in their responses; while some took on the responsibilities of prefectship, house-captaincy, etc., many remained isolated and disaffected. In general the picture showed that as children they did not find it at all easy to make friends, or find peers with similar interests to their own. This was illustrated by the following descriptions of their school lives:

> 'I hated school, especially all the social bits of it. I spent my time in primary school keeping silkworms and frogs, climbing and walking—things on my own I suppose. I was more serious than other children, introspective, solitary. I just had one or two friends that I stuck to. I was very shy, I felt very lonely when I was young—now I enjoy it. Boarding school was even worse, I hated this even more. I did go through a bad phase then, I was against everything and everybody.'

> 'On the whole school could have been worse, but I didn't enjoy it, I was bored. I never wanted to learn what everyone else was learning—I used an encyclopaedia, the teachers were awful—I taught myself to read, and read avidly. Most of the time I was alone. I was very shy from an early age. I was very naughty and fat, I just played the clown. I was terribly lonely as a child, I associated with children much older than me. I always felt special and rather odd.'

> 'I didn't enjoy school when I was young, I was bullied a lot. I spent my time playing the cello, with electronics and crystal sets and doing geometry for fun. I think I was very shy, I felt excluded from the social group of the class. Things got better after 16, then I enjoyed it a lot, it was a different school.'

> 'It's difficult to remember. My outlook on life was different from others. I felt different, more mature, but in a way that wasn't true. If I had been more mature I would have known how to get on with others better. I was alone and solitary.'

> 'I loved it, especially the sciences. I had a small circle of friends, no enemies. I was very shy from childhood, introverted, I'm a bit better now. I had a free place at school, I felt different, I didn't seem to have the same hobbies, I spent all my spare time with radios and electronics from five years old. I sometimes felt superior. I was above average, I used to do other people's homework.'

These quotations confirmed the Dependents early interest in the more esoteric, solitary activities and hobbies, but the most significant description they gave of themselves was their inability to mix well with others from an early age and their feelings of extreme shyness. Of the 45 Dependents who were interviewed, 32 described themselves as solitary children. Not only did they feel merely shy, but different and more serious than their peers with whom they felt unable to mix. Many children are shy, and many have at some time or another been ostracized by others, especially if considered to be highly intelligent or different in some way. As the three groups had been matched for highest educational level, any differences could not be attributed to academic abilities alone, but could reflect the different biases which controlled their interactions and interests.

Adult social experiences

The interview data indicated that the Dependents' tertiary level education, which for most was within a university setting, seemed to provide a happier, more stimulating environment than they found at school. Although most admitted still to being extremely shy, there were more opportunities for them to interact with others who shared their interests. Nevertheless, their social lives tended still to be confined to two or three like-minded friends, and many remained rather solitary, preferring to spend much of their time alone with their studies or on solitary activities. Some became members of societies which reflected their preferred hobbies; joining the mechanical, scientific or technological societies, as well as computer, radio, science fiction and *Dungeons and Dragons* clubs.

This period of early adulthood, whether studying or working, was the time when most seriously wished to develop their relationships with the opposite sex. Of the few women interviewed none appeared to find great difficulty in forming partnerships and all had married, although two of the four had married very young in opposition to their parents' wishes and another was 'happily divorced'. However, for quite a large proportion of the men finding partners was not easy.

> 'They were not interested in the things I was. They were not interested in me. I never worked with any females, although at university 20 per cent were girls they were not from similar backgrounds and so were unsuitable for me.'
>
> 'I never found it easy to make friends, my head responds faster than my heart. I was a late developer in that sense, although I did get close to marriage once.'
>
> 'It's difficult to find people. They'd have to be intelligent—interested in computers a plus. I'd never initiate, I'd wait to be chatted up.'
>
> 'It was always very difficult. Most people find it difficult. Men and women are interested in different sorts of things, naturally. I didn't have the right social skills, but I always did the chasing.'
>
> 'Basically girls aren't serious enough. It's difficult to find people old enough to cope with me.'
>
> 'Perhaps I've denied myself the opportunities. Up until now I've been happy at work and with my computers, it's easier than going out—it's safer. I'm too shy, it's risky.'

Many of the men described themselves as 'late developers' and a few, well into their twenties and thirties, had not had a serious relationship although the majority did aspire to marriage. Although most appeared to enjoy the company of women, often in preference to their own sex whom they often found boorish, the found it difficult to overcome their shyness so that relationships could develop. For those who had found partners, few seemed to have been the initiator of the relationship, most having relied upon the woman to make her feelings known. Most reported that their parents had told them nothing about sexual matters, and this may have proved to be an inhibiting factor.

Although all appeared to want a true partnership, a sharing of minds, they had found it difficult to meet women who exhibited the same interests as themselves. Because of the preponderance of scientists among the group, few studied subjects which were as equally popular with females, and many of their club and society activities appeared to be male-oriented and therefore did not provide an avenue to meet women. Those who had married often seemed to have married their first or second partner and settled down very quickly. A few of the men seemed only to be interested in older women, and two of those interviewed had married women more than 12 years older than themselves. This picture was not true for all of the respondents, as a few found in women the companionship that they had been unable to develop with members of their own sex, and some had had many and varied relationships. More than half of the interviewed Dependents did marry, two during the course of this research. Most seemed to be very content with their marriages, finding fulfilment from the relationship with their partners and from parenthood.

> 'It's the ability to have someone to discuss things with very intimately. When you're alone you have no-one to share with. The responsibility makes you strive harder for advancement, you may give up on your own—you try to achieve more, a better standard of living. It's having someone you can come home to and feel right with—you don't have to worry about feeling lonely. However the advantages are also the disadvantages. The responsibilities mean you have to achieve and have to come home.'

> 'Having someone to plan and share with. I do more ambitious things now, like buying this house. It's the general mind-stretching effects of living with someone with different interests. The biggest disadvantage is the lack of freedom.'

> 'There is more than one of you to face any problems. Someone to help you through the bad times—both ways, and the advantages of children.'

The majority of the interviewees truly valued the benefits they had gained from the close companionship of marriage. The partnership had removed much of the tension and loneliness of their earlier years and had given them stability and purpose. Almost all regretted the loss of freedom to do exactly as they wished but this seemed more than made up for by the advantages. Not all of the marriages were unsullied however, and for some the advent of the computer had caused problems within their relationships (as discussed in Chapter 10.) In spite of the greater satisfaction found during adult life many still did not find it easy to make friends and rarely initiated social interaction. However, whatever the difficulties still experienced most seemed to be far more content with their lives than when they were younger. This was not only due to their computing and their families, but also to their jobs.

Work experiences

All but two of the 45 interviewees were either in employment or studying. Their work experiences differed greatly, from the few who had remained in the same profession with the same organization from the start of their careers to those who changed posts frequently. While some were determined to further their careers by achieving regular promotions, quite a large proportion appeared totally disinterested in self-advancement of this type. Although some aspired to higher incomes, to many increased prestige and status seemed relatively unimportant; they wished rather to enjoy their work without the additional stress of extra responsibilities and authority which attends those in the highest positions.

All but three, two of whom were women, had had backgrounds of work experience which were scientific or technologically based, if not directly concerned with computing from the first. Even two of the three schoolboys interviewed had already had programs published and one, following in his father's footsteps, was a contract programmer during the holidays. Five were teachers of computing, either within schools or universities, all of whom had come from backgrounds in either mathematics or science. Two-thirds of those interviewed worked directly with computers, and 10 of the 45 interviewees had been the main initiators of computing within their work setting.

Many felt themselves to be in their ideal job, having occupations which made full use of their talents. Although most were not as interested in the computing they had to undertake at work, they still gained great fulfilment from their jobs. Five were self-employed and others aspired to this status, although a few would have preferred not to have worked at all, seeing the ideal as a life which allowed them to explore their computing interests unhindered. Those who showed any dissatisfaction with their jobs tended to be those in employment which had not been computerized.

Comparisons between the adult experiences of the groups

Multi-choice questions addressed some of the issues covered in the interviews, and Table 8.3 summarizes the results from the three groups.

Few significant differences were observed. The Dependents were found to be no more likely to have been 'late developers' than the control groups, and the majority of all three groups had had partners. Few considered themselves to have been the initiator of relationships or had been told the 'facts of life' by their parents. However, the Dependents were significantly more likely to have partners older than themselves, with 24 per cent of the Dependents but only 10 per cent of the control groups reporting them to be older.

Table 8.3 Adult social and work experiences

Questions on adult experiences	Significance	Comments
Age started dating	–	Average age 16·3, S.D. 2·8 years
Have had partners	–	Only seven or eight of each group had not
Initiator in relationships	–	Only one-third felt they were
Parents told 'facts of life'	–	Two-thirds were told nothing by parents
Age of partners	*	Dependents partners were older
Easy to make friends now	***	Dependents more difficult
Contented with life	–	Majority were contented
Main fulfilment in life:		
–work	–	One-third from each group
–family/friends	**	Dependents less
–hobbies	***	Dependents more
Computers used at work	–	Two-thirds had used them at work/study
Career advancement important	–	Unimportant to one-third
Job makes use of talents	–	Half said it did
Job controlled by others	**	Dependents had less autonomy
Work for enjoyment/necessity	*	Dependents work for enjoyment
Differentiate between work/play	**	Dependents less likely to

Even in their adult years the Dependents were still less likely to socialize than either of the control groups. The Dependents spent far fewer evenings socializing, as has been previously discussed, and 43 per cent of the Dependent group stated that they still found it difficult to make friends, in comparison with only 19 per cent of the control groups.

The majority of all three groups stated that they felt contented with their lives, but there were significant differences between them when asked from which aspect of life they gained their greatest fulfilment. (Table 8.4 summarizes their responses). Not all of the respondents were able to make merely one choice between those offered, as can be seen by the total number of responses. When each set of responses was analyzed between the numbers who had and had not selected each choice, the results showed that there were no significant differences between the groups for the numbers who gained their main fulfilment from their work, but significant differences were observed between the Dependents and the controls for the numbers who selected their families and friends and those who had selected their hobbies as their main source of fulfilment. While 59 per cent of the Dependents gained their greatest fulfilment from their hobbies, only 27 per cent of the controls did, but a reverse trend was shown for those citing their families and friends at 32 per cent for the Dependents and 53 per cent for the controls.

Table 8.4 Source of main fulfilment in life ($N = 75$ in each group)

Main fulfilment in life	Dependents	Owners	Non-owners	p
Work	22	25	27	–
Family and/or friends	24	38	41	**
Hobby, including computing	44	18	23	***
TOTALS	90	81	91	

In their working lives, no differences were found between the groups for those for whom a career structure with advancement was felt to be important, and the majority of all three groups felt themselves to be in jobs which used their talents to the full. However, the Dependents were less likely to have complete autonomy in their jobs than the control groups. Although the Dependents had work histories showing steady promotion throughout their careers, these tended not to be to managerial positions, which could account for this lack of autonomy.

Far more of the Dependents than the controls worked for enjoyment rather than necessity. This was further clarified when significantly more of the Dependents admitted that they found it difficult to differentiate between work and play. The results from the Leisure Activities Inventory had indicated that their hobbies, which tended to be technologically and scientifically biased, differed from those of the controls (who were more likely to include sporting and social activities in their preferred leisure pursuits). If one's job almost mirrors the activities in which one would choose to indulge in one's spare time, it is possible to see that for the Dependents work would merely be an extension of what was also seen as play; an ideal situation to which most would aspire.

The importance of social contact

As the Dependents considered themselves to be significantly more serious and shy than the controls and different from their peers, such feelings of uniqueness and isolation while young could have compounded their need to find solace in activities which did not require the participation of others. It was significant that it was the primary school years in particular that the Dependents did not enjoy. During these formative years much school work concentrates upon practical, social and cooperative activities in order to aid the development of the integrated personality, while work at the secondary level centres more upon logical reasoning, academic success and examination subjects. That the Dependents enjoyed their secondary schooling as much as the control groups did not necessarily imply that their social difficulties had been overcome, but rather that they no longer caused such distress. The interviews revealed that the secondary school biases more closely mirrored the Dependents' own interests, allowed them to specialize in the technological and science subjects, and as a consequence enabled them to gain recognition for their abilities, thereby increasing their perceived status and worth. In addition, they appeared to have come to terms with their lack of social skills and tended to make friends only with a small number of like-minded pupils from whom they gained reassurance.

Although it is always more difficult to establish one's identity in an environment closely controlled by others, as in school where one has little autonomy or power, it was felt that during the years of tertiary education and work experience that the Dependents' results would more closely mirror those of the control groups. However the results persistently pointed to the fact that the most significant difference between the Dependents and the control groups lay in their lack of ability to form social relationships. Even during adulthood the Dependents were shown to lead more solitary, isolated lives than the controls and to depend heavily upon work and leisure activities for their satisfaction, activities which were object- rather than people-centred.

Those Dependents who had married, and these were of similar proportion as in the control groups, had for the most part been able to escape from the feelings of loneliness which had dogged most of them throughout life. That most appear to have married their first or second partner was in accord with the research of Leary and Dobbins (1983) who, perhaps not surprisingly, found shy people to have fewer sexual partners and less experience. However, without the relevant, comparative data it was not known whether the Dependents differed in this respect from the controls. Using the Dependents' criteria, females on the whole were considered to be immature, giggly and neither serious nor old enough to act as adequate partners for them. It was therefore significant that quite a number of the Dependents, male as well as female, had chosen partners who were much older than themselves. One could perhaps consider that they were seeking the physical contact and warmth denied them by their parents.

Although marriage does fend off isolation and loneliness, Bowlby (1973) did not necessarily believe that this led to greater happiness or satisfaction for lonely people. The results confirmed this to some extent. Unlike the majority of the controls, for whom family and friends were paramount, the Dependents gained their greatest contentment and satisfaction from their computing, perhaps indicating that they were still unable to share themselves fully with another person (in spite of their expressions of satisfaction with their marriages). Many appeared to have very little in the way of common interests with their partners, and found general conversation and sharing difficult. This was confirmed by the partners of the Dependents, as is discussed in Chapter10.

Having interests which were object-centred and not socially oriented, and with many reporting themselves to be lonely as well as shy, it was perhaps not surprising that the male Dependents in particular found it difficult to converse with women, as was shown in their quotations. Not only this, but many found difficulties with male-male conversations because of lack of common interests with most of their peers.

Jourard (1971) observed that the inability to disclose intimate details of oneself to others was closely correlated with loneliness. Similarly, Solano *et al.* (1982) found that lonely people were significantly less likely to make themselves known to other people, preferring to discuss impersonal topics, and that a lack of ability to talk about their personalities, particularly by men in conversation with women, was correlated with the greatest feelings of loneliness. (For women loneliness was associated with the inability to disclose personal information to either sex).

Such a lack of reciprocal, intimate exchange inhibits the normal development of social relationships, and Altman and Taylor (1973) found that lonely people were unable to perceive this lack of intimacy in their conversation and therefore were unable to understand their inability to form lasting friendships. In addition, Jones *et al.* (1981) found that lonely people tended not to like or trust other people, a factor which therefore inhibits intimate exchange and perpetuates the feeling of isolation. This in turn leads to further rejection, to which the lonely have been found to react very negatively to the extent of becoming overtly and verbally aggressive, behaviour which has been observed by Check *et al.* (1985) to be especially true from men towards women. I was not made aware, by either the interviewees or their spouses, that aggressive behaviour was manifest within relationships, but mere retreat into the computer room may have been a coping strategy used to avoid such confrontations.

During the interviews it became apparent that the Dependents frequently found it very difficult to talk about themselves and their parents, some seeing no relevance to this line of questioning. However, trust was established and a satisfactory interchange developed because of our mutual interests in computers and computer dependency, subjects of paramount importance to them but of little interest to most of their acquaintances and family members. Many were very unfamiliar with disclosing details about themselves and often seemed unable to describe themselves or members of their family in any detail. It seemed less that they were inhibited or unwilling to talk but more that they were unused to such intimate conversations. It was not unusual for interviewees to state that they had never discussed these subjects with others before, not even with their spouses. However, although almost all the interviewees reported themselves to be very shy, this was not particularly apparent during the interviews where the main topic was of mutual interest to both parties. This was understandable, as the interviewer-directed conversation eliminated the need for them to take the initiative in the interaction, found by Pilkonis (1977) to be one of the major manifestations of shyness, and in addition the task-related aspects of the interview eliminated the shy person's anxiety found to be present within unfocused interpersonal encounters. The interview situation was therefore atypical of the normal social interactions they would encounter.

Shyness seems difficult to define. It may show the characteristics of reticence and inhibition (Powell, 1981), garrulousness and assertiveness (Litwinski, 1950) or aggressive tendencies (Kaplan, 1972). Whatever its characteristics, Zimbardo (1977) conceded that 'ultimately you are shy if you *think* you are, regardless of how you act in public'. He also held that the transitory nature of modern life, with its high mobility and lack of an extended family, perpetuated the alienation of one person from another, leading people to be mistrustful of forming long-term relationships (Zimbardo, 1980) and that shy people chose the security of isolation in preference to the risk of being rejected by others (Zimbardo, 1981).

Harris (1984) believed shyness to be characterized by anxiety and self-consciousness, and saw it to be socially rather than psychologically induced, which would accord with the research dealing with the development of the object-centred bias. Smail (1984) saw shyness as a manifestation of the sensitive person's awareness of their inadequacies, and concluded that perhaps 'shy people are close to the recognition of a truth which their more confident fellows have more successfully repressed'. He believed this self-consciousness to be caused by parental influence, especially that of the middle classes who use expectations of success as a means of manipulating affection and love. Similarly, Weiss (1962) believed that lack of emotional and physical closeness between parent and child, over-protective mothering, or love conditional upon performance, leads to the development of safe behaviours. These can centre upon tasks which have to be performed to a high standard in order to try to gain approval or to subjugation of the emotions by learning not to feel and care. Both of these seemed to be present, to a greater or lesser extent, within the Dependents coping mechanisms.

Such theories bear very close resemblance to those of Roe (1957) and Hudson (1966) who saw object-centred biases and convergence as effective means of coping with life, stemming from an awareness of one's limitations and strengths, and not as a neurotic, unhealthy reaction. In addition, they ascribed these biases, as with the theories of shyness, to the task-centred relationship between parent and child during the early years of life. Both object-centred bias and shyness seem to develop simultaneously from the same source and to be inextricably linked with each other.

Many people consider themselves to be shy, as did some of the control group members, but not all become dependent upon an activity such as computing. No doubt their needs and coping strategies differ. Severe difficulties experienced in social situations do however help to explain the satisfaction found when interacting with a computer, as such an interaction gives the semblance of a real and meaningful relationship. To investigate the social and psychological aspects more fully, an in-depth study of the personalities of the Dependents was undertaken in order to determine which traits and factors were apparent within the Dependent group which differentiated them from the controls and the general population.

Personality factors

As the previous results suggested considerable differences exist between the interests, cognitive styles and social abilities of the Dependents and the control groups, the aim of this section of the research was to investigate whether these differences would be apparent when measurements of personality traits and factors were examined.

Initally, the 45 interviewed Dependents were asked to make self-evaluations of their own personalities, and in addition were asked to state which characteristics they would like to have which were perhaps lacking. From the interview data, two scales were devised from bi-polar adjectives drawn from the self-descriptions of the Dependents in order to make comparisons between the Dependents and the control groups; the *Self Report Scale* and the *Ideal Self Scale.* Each respondent was asked to give descriptions of how they truly saw themselves on the Self Report Scale, and was also asked to complete the Ideal Self Scale to examine differences in perceived factors considered lacking in their own personalities.

Although standardized pyschometric tests could not be used to make direct comparisons between the Dependents and the controls (ethical constraints prevent them from being sent for self-administration), tests of personality were given to the 45 interviewed Dependents, allowing them to be compared with population norms. Batteries of univariate tests were inappropriate as this research aimed to be exploratory, therefore a multi-dimensional personality test was selected; Cattell's *Sixteen Personality Factor Questionnaire* (16PF). The *Group Embedded Figures Test* GEFT, Oltman *et al.*, 1971) was also completed by the majority of the interviewed Dependents. This involves the discrimination of simple shapes from within complex figures, and its successful completion has been correlated with field-independence; the ability to concentrate upon the particular while ignoring the 'noise'.

Self Reports

Many of the Dependents found the questions about their own personalities very difficult to answer, and in comparison with their descriptions about computers often became hesitant and monosyllabic. None gave the impression that they did not wish to disclose this information, but rather that they had never before attempted to evaluate themselves in this manner and had little idea of what type of personality they had. Most concentrated upon certain aspects of their social lives, obviously a very important aspect to them, but few mentioned specific personality traits or their emotional states. A large number of quotations have been included in this section to highlight the similarities between the Dependents and to illustrate the ways in which they viewed themselves and conducted their lives. Most of the following include the complete statements made about themselves, indicating the brevity of their responses.

> 'I couldn't given an impression of myself, absolutely note. I'm shy. I'm one of the boring, average people. I'd like to be more extrovert, more commanding, I'd like to have more social confidence.'

> 'Dear me, I find that very difficult. Workwise I enjoy being a leader and knowing my job. At work I'm seen as incomprehensible to some—I have ideas which are ahead of others, far-fetched. I'm a good Dad and a fair husband. I'm reluctant to analyze myself.'

> 'I think I'm a leader—I can't bear the fuddyduddying of everyone else. I can't sit back if something needs saying—I'm bossy. I'm as hard as nails, for years I couldn't cry. Being divorced has hardened me—I can't bear most men. I don't even kiss my own children—I don't display emotion.'

> 'I'm a miserable old cuss, that sums it up. I find it difficult to answer these introspective questions.'

> 'I'm a loner, not as socially outgoing as I'd like to be, shyer. I'm a little eccentric, I have unusual interests, I don't react with a crowd, I'm a loner.'

> 'I'm wasted, talent-wise. I've missed out on a lot and am trying to patch it up. I try not to be too introverted, I try not to get into it. I'm eccentric and unusual.'

> 'I'm an introvert, intelligent, not at all gregarious and rather neurotic.'

> 'Impulsive, disorganized, sometimes not very nice, cynical about other people, opinionated.'

> 'I'm different—totally honest, very bad at keeping secrets, easily influenced, non-competitive compared with others, not very modest. "Square", people used to call me that a lot—I didn't mind. I'm tolerant, intelligent, responsible. I'm ambitious and inquisitive. I dislike being bored by other people and their petty attitudes.'

> 'I'm unconventional, in my opinion. I'm fairly easy going. I spend a lot of time asleep these days—I get bored very easily with life. I get along with anybody at first then difficulties come. I'm fairly introverted.'

'I wouldn't say I was hardworking, except for things that interest me. I may not be as sociable as I could be. Others chat, I go to one side and write down a program. I'm unsociable and a bit of an idiot about computing.'

'I'm impatient and suffer fools badly. I am in the correct terminology introverted and therefore tend to be solitary. I'd like to be able to judge people better.'

'It's very difficult to say. Quiet, somewhat introverted, a bit rigid in my views and attitudes. I'm searching for something. I have the potential and the abilities to do amazing things, but there are severe blocks especially with relationships, trying to be friendly, etc. I'm quiet and introverted but also the Williams schoolboy—into hacking and dreadful tricks, a Peter Pan.'

'I'm a very "conscious" person—a lot of my life is governed by conscious decisions rather than emotional reaction. I'm non-competitive and non-ambitious. I'm eccentric—I may not seem friendly, but I think I am.'

'I'm incredibly average, but I know I'm not really. I'm a cut above intellectually. I have got beyond the point of wanting to be everybody's friend. I'm contented with my own company, I can spend a lot of time alone and I'm rarely bored. I used to get depressed a lot, but I get a lot of "highs". I think people see me as a bit of a loony, because I outwardly seem to be obsessed with things to the exclusion of other activities. Out of a pound. I spend 80 pence on computing, 10 pence on food, and 10 pence on rent.'

'Irreverent about everything—I don't take anything seriously. I'm fairly intelligent, not a socialite or sporty. Self-critical but unwilling to do anything about it. I'm shy and introverted. I don't suffer fools gladly but I'm lackadaisical, placid, easy-going.'

Confirming previous results, the overwhelming majority of the interviewees described themselves as shy, insular and solitary people. Not all seemed unhappy with this state, however, and quite a number stated that they had little need of others, felt themselves to be self-reliant and were perfectly content with their own company. Their perceived lack of basic communication skills, coupled with an interest in computers not shared with many others, left some feeling cynical and opinionated. Some merely described themselves as average and ordinary but many saw themselves as intellectually superior, inventive and inquisitive types of people who suffered fools badly. Of those who would have liked to change their personalities, most concluded that they would enjoy having greater social abilities in order to reduce their feelings of isolation and difference from others. Although many quite forceably stated that they did not care about the opinions of others, not wanting to be like the masses, many did want to be liked and wished they had more confidence to share with others effectively.

When asked how they thought others saw them, the Dependents' answers reflected their own descriptions, although often they were more vehement and self-denigrating. Some thought they were seen as 'weird', 'eccentric' 'loonys', unable to join in the more usual activities of life with any degree of comfort and ease, although some obviously felt proud of this distinction from others.

Despite their self-descriptions, which were on the whole derogatory and denigrating, I found them to be extraordinarily hospitable, stimulating, amusing and intelligent conversationalists. Although describing themselves as having poor social skills, most were not self-centred in their conversations. They were interested in my work, my own history and my desire to study the area, and many wanted to learn whether the other interviewees were similar to themselves. This seemed especially important to those without modems, who had not had the chance to communicate with like-minds over the networks.

These data gave a very good impression of the interviewees; however in order to discover whether their self-descriptions held true for the whole of the Dependent group and whether they differed in any way from the two control groups who had been matched with them, all three groups completed the Self Report and the Ideal Self Scales.

Results from the Self Report and the Ideal Self Scales

Significant differences between the groups were found in seven instances on the Self Report Scale, all of which showed these differences to lie between the Dependents and the control groups. These seven significant differences strongly affirm the Dependents' self-descriptions, and are illustrated by their mean scores in Figure 8.4. They have been reversed in some cases and reordered according to the mean scores of the Dependents, with the mean scores of the two control groups combined.

The Dependents considered themselves to be isolated, introverted, shy and misunderstood, while the controls showed themselves to be extroverted, outgoing, understood and not isolated. The Dependents also rated themselves as more curious and materialistic and less attractive than did the the controls.

The results from the Ideal Self Scale showed no significant differences between the three groups for any of the bi-polar scales. This demonstrates that all three groups of respondents shared the same opinions as to which characteristics they considered to be ideal.

When comparisons were made between the Self Report and the Ideal Self Scales, the results from the bi-polar measures indicated that in most instances all three groups wished to change certain aspects of their personalities. The data showing differences between the two scales have been sub-divided for clarity, and the first section contains the seven results which already showed differences between the Dependents and the control groups for their responses to the Self Scale (see Figure 8.5).

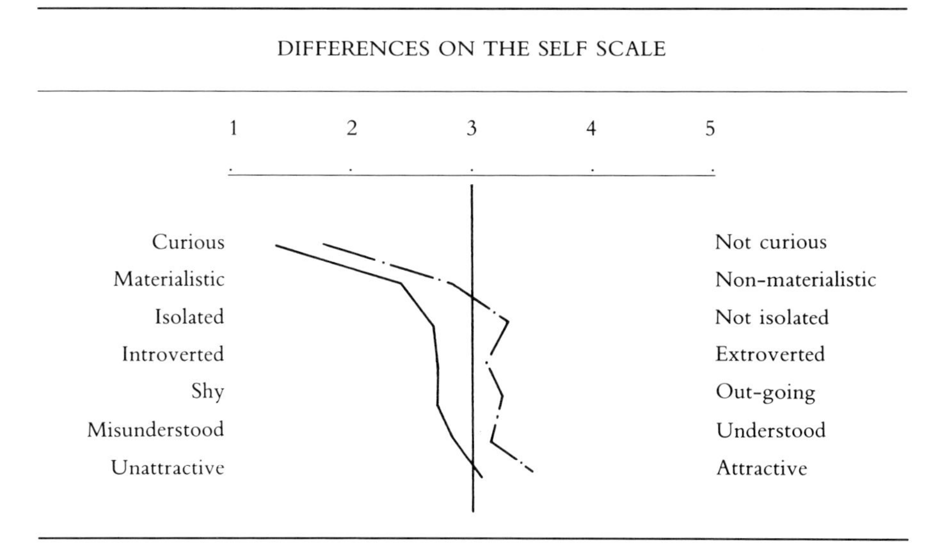

Figure 8.4 Profiles of differences on the Self Report Scale
Dependents ______ , Control groups .__.__.

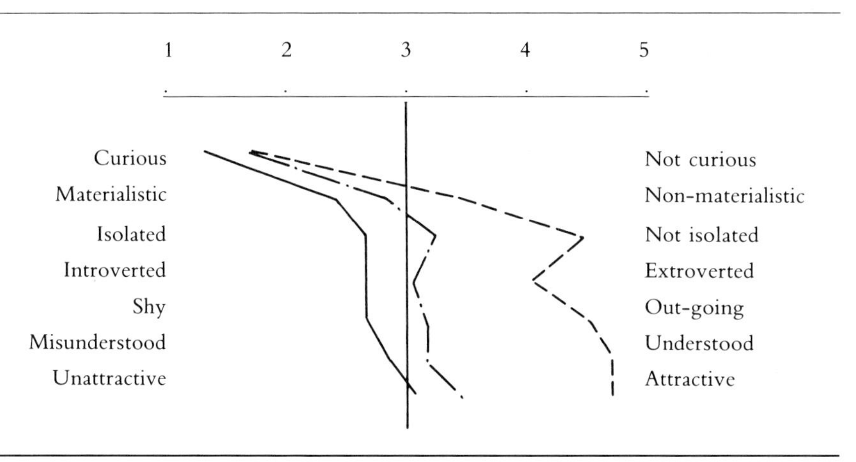

Figure 8.5 Profile of differences between the Self Report and Ideal Self scales, and between the Dependents and control groups

Self Report: Dependents ______ , Control groups .__.__.

Ideal Self: All groups _ _ _ _

While the control groups wished ideally to be more curious than they felt they were, no significant difference was observed for the Dependents between the Self Report and Ideal Self Scales for this variable. The remaining results showed that although all three groups wished to improve their social skills somewhat and wished to be less materialistic and more attractive, the results from the Dependents showed greater disparity between self and ideal self than did the results from the control groups.

The second sub-section contained the results which showed no significant differences between the three groups of respondents but did show significant differences between the Self Report and the Ideal Self Scales. These data are illustrated in Figure 8.6, and the bi-polar scales have been reordered, reversed where necessary, and divided into logical categories. In this figure only the combined mean scores from all three groups are illustrated.

Assuming that the respondents considered their ideal attributes to be 'positive', it was apparent that all three groups felt that their general state of mind was not as positive or as satisfactory as they would have wished. Although all of their self-ratings already lay on the 'positive' side of these variables, they wished their ideal selves to have more of these characteristics. Similarly they wished to improve their attitudes towards others and to be more stable emotionally. Also, despite showing that they considered themselves to be very intelligent, imaginative and creative, all wished ideally to have the maximum quota of these attributes.

On only six occasions were no significant differences observed between the results from the Self Report scale and the Ideal Self scale nor between the three groups. These showed that the respondents had already obtained their desired, ideal characteristics and had no wish to change further. These are displayed in Figure 8.7, with the mean results from the three groups combined and the results from the two scales running in close parallel.

One may observe from the results of the Self Report and Ideal Self scales that on the whole these three matched groups of respondents, most of whom were young men, offered very similar descriptions of their own personalities and their ideals, and that the main differences between the groups were those which reflected the social difficulties reported by the Dependents.

Results from the Group Embedded Figures Test (GEFT)

The *Group Embedded Figures Test* was given only to the interviewed Dependents, and the results confirmed that the Dependents as a group could be considered to be field-independent as measured by this test. Comparisons made between the Dependents and the adult norms for the GEFT (Oltman *et al.*, 1971), showed that 79 per cent of the group produced results which lay in the upper quartile of the normal range. (i.e., results expected to be achieved by only 25 per cent of the population.)

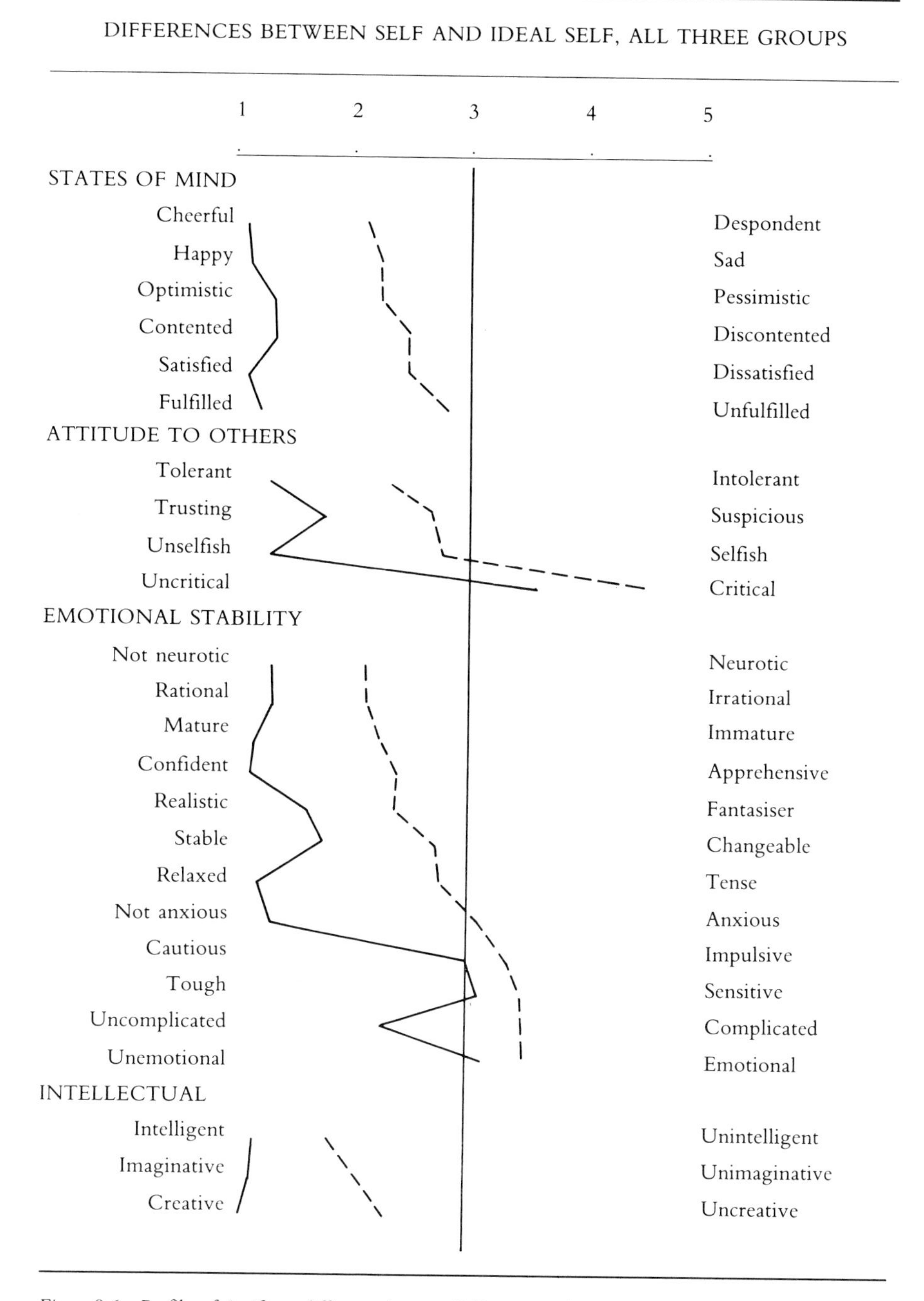

Figure 8.6 Profiles of significant differences between Self Report and Ideal Self scales, no significant differences between groups

Self Report _ _ _ _ , Ideal Self ______ , (all groups)

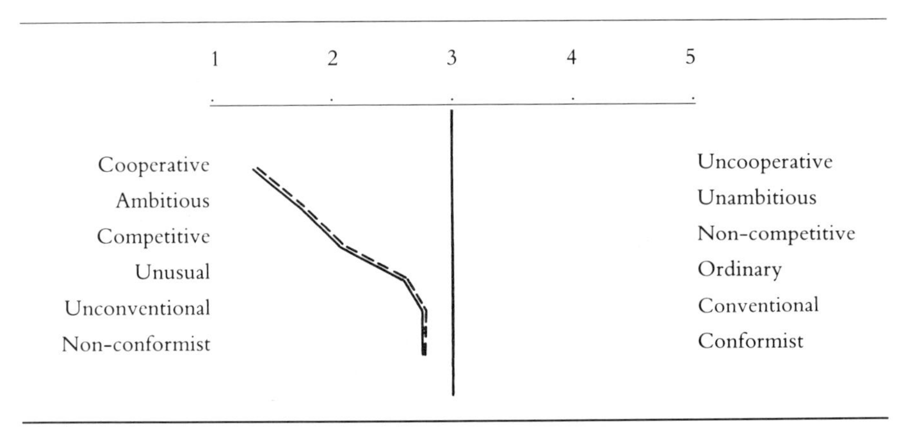

Figure 8.7 Profile of results showing no significant differences between the Self Report and Ideal Self scales, nor between the groups

Self Report ______ , Ideal Self _ _ _ _ , (all groups combined)

Table 8.5 gives the adult norms, that is the total number of correct responses for each of the quartile ranges according to the sex of the testee. The results showed that for the 18 figures of the test the Dependents yielded a mean of 16·0 correct responses with a standard deviation of 2·6; scores higher than the norm and with a smaller variance.

Although allowed 10 minutes to complete the test (5 minutes for each half) the mean finishing time for the group was 7 minutes 38 seconds, nearly 2 minutes 30 seconds faster than normally allowed. The fastest

Table 8.5 Comparison of the Dependents' GEFT scores with the adult quartile norms, plus means and S.D., by sex

Quartiles	Male norms Qs. correct	Male Dependents Freq.	Female norms Qs. correct	Female Dependents Freq.	All Dependents Totals
1	0–9	1	0–8	0	1
2	10–12	2	9–11	1	3
3	13–15	3	12–14	0	3
4	16–18	24	15–18	3	27
N	155	30	242	4	34
Mean	12·0	16·2	10·8	14·3	16·0
S.D.	4·1	2·6	4·2	2·6	2·6

person to complete all 18 questions correctly completed both halves of the test in just under 4 minutes.

Two facets of the Dependents' personalities were reflected during this course of testing. When faced with a figure which they could not immediately distinguish from its background many would doggedly persevere, ignoring those items which were left. Consequently seven people ran out of time for one part of the test, and three for both. For these 13 people the average number of correct figures was only 5·3 of the nine possible, while the group mean was 8·0 for each half. Their very tenacity prevented them from completing the test and illustrated their determination not to be outwitted by a single item. The second aspect involved those who were able to complete the test within the time limit. Of them, only two people checked their answers, even though all were told the length of time which was outstanding. This not only applied to the 11 people who had achieved perfect results but also to those who had made mistakes. They appeared to recognize their abilities, saw no immediate problems with this type of test, and seemed convinced they had completed all of the test items perfectly.

The results from the GEFT strongly supported the deductions made in Chapter 7 concerning the Dependents' preferred leisure pursuits and showed that the interviewees were very heavily biased towards a field-independent cognitive style as measured by this test. This has been related to other areas of analytical intellectual ability where symbolic representations are dealt with. Field-independence is also associated with a developed sense of separate identity and selfhood, among people who are unlikely to revise their views because of social influences and attitudes (Bell, 1955). In times of stress such people tend to use the specific defence mechanism of isolation from others in order to separate feelings from thoughts and ideas (Witkin *et al.*, 1962), and at the cost of an emotional life place great reliance upon intellectualization. In addition, field-independence has been thought to develop in childhood within families where separate, autonomous functioning has been encouraged (Barclay and Cusumano, 1967).

Such results are consistent with the Dependents' descriptions of their own personalities and family lives. Elliot (1961) summarized the self-ratings of the field-independent as a 'touchy, brooding, intellectual malcontent, some solitary mixture of prophet, beatnik and angry young man'; a description of which I feel sure some Dependents would be proud.

Results from the Sixteen Personality Factor Questionnaire (16PF)

The *16PF* was completed by the 45 Dependent interviewees only. The conversion of their 16PF raw scores to sten scores was undertaken using the British General Adult Population Norm tables (Saville, 1972), using the male or female norms for Form A as appropriate. Second Order factors were calculated using the procedures laid down by IPAT (1979).

The results showed that for six of the twenty First and Second Order factors the Dependents were significantly different from the adult norms, with sten scores of more than one standard deviation from 5·5 sten (see Figure 8.8). The Dependents were shown to be of above average intelligence and were also highly assertive, imaginative and self-sufficient, and in addition yielded high mean scores for tough poise and independence.

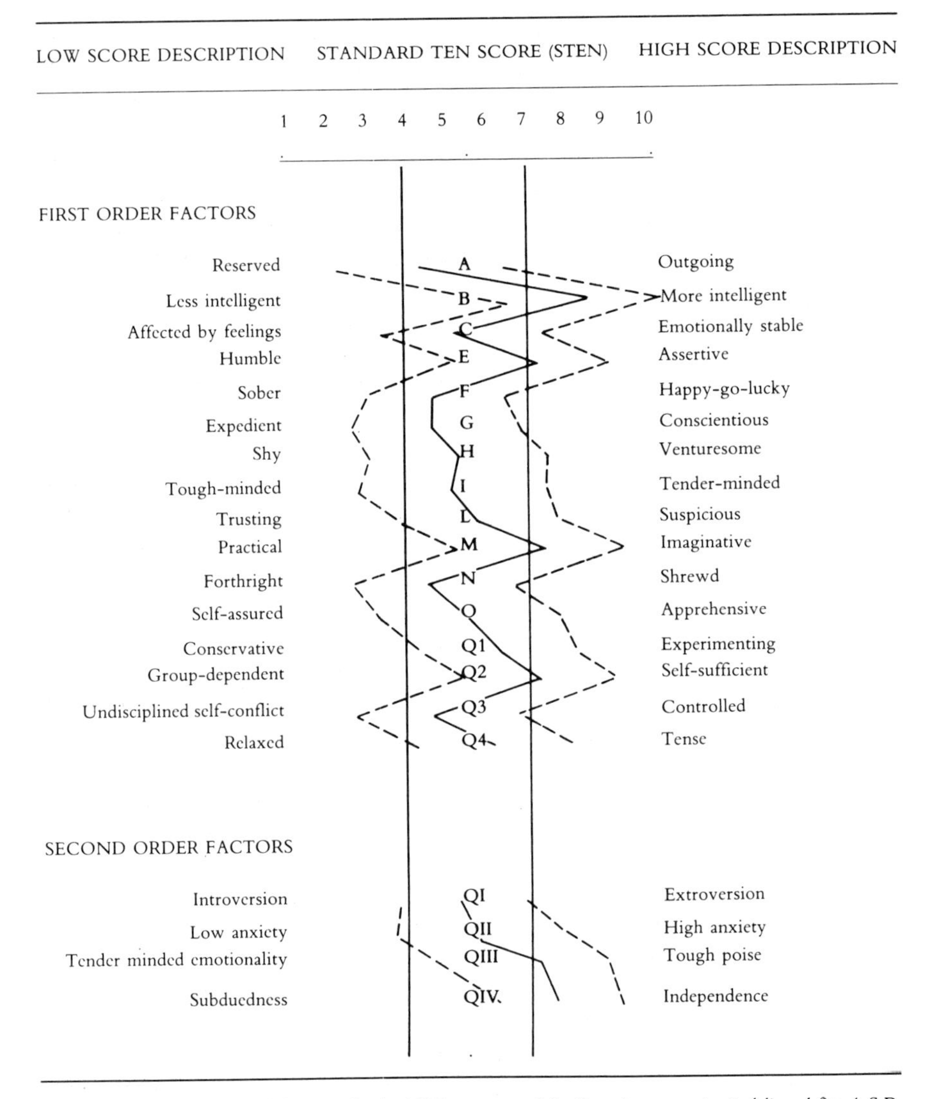

Figure 8.8 Means and standard deviations for the 16PF sten scores of the Dependent group (vertical lines define 1 S.D. from the population norm).

There was wide variation in the results from the Dependent group and none of the factors showed a standard deviation of less than 1·0 (as illustrated in Figure 8.8). Although these deviations were frequently smaller than those given by Cattell *et al.*, (1970) for results from much smaller groups, which for many years have been used for vocational, educational or diagnostic purposes, a further analysis was carried out. This used the model devised by Krug (1981), who drew profiles from modal rather than mean scores, believing them to give a more realistic picture of group results.

Using Krug's model, the frequency of the sten score for each factor was determined in order that the modal values could be isolated. This enabled observation of the true variations within the group to be carried out.

The majority of the factors gave uni-modal distributions about the group means, and as such can therefore be assumed truly to represent the group's results. However, six of the First Order factors showed bi-modal distributions which deviated not only about the group mean but in five cases also deviated about the general population means. The profiles of the frequencies of each sten score for these factors (H, I, M, O, Q1 and Q3) are illustrated in Figure 8.9. The results suggested that the Dependents did not form a coherent whole but that there were different types of personality within the group.

The largest deviation from the population mean occurred for factor B, not a fundamental measure of personality but a measure of intelligence. The modal score for the group was 10, the maximum possible, and was in keeping with the educational attainment and employment of the Dependents, quite a number of whom were members of *mensa.*

Nearly two-thirds of the group (62 per cent) also showed high scores for Q1 (conservative : experimenting), results consistent with those who are primarily interested in intellectual activities, are open to innovation and not adverse to change. High scorers have been shown to be well-informed, problem-solving and questioning, often scientifically-oriented and preferring self-teaching to classroom instruction; all aspects which were in keeping with the Dependents' interest in new technology. [All interpretations are taken from the Handbook for the 16PF (Cattell *et al.*, 1970) or from the Manual (IPAT, 1979) unless otherwise stated].

The Dependents yielded very high scores for factor M (practical : imaginative), and as such are typified as being self-motivated, imaginative, even absent-minded in outlook. High scorers are prone to fantasize and are absorbed in ideas, often being oblivious to everyday practicalities; a fact confirmed not only by the Dependents but also by their spouses. The extreme individuality of such high scorers also tends to see them rejected in group activities, and such scores are often seen in male engineers (Roe and Siegelman, 1964). Once again these results corroborate the interview data, and although the results were bi-modal

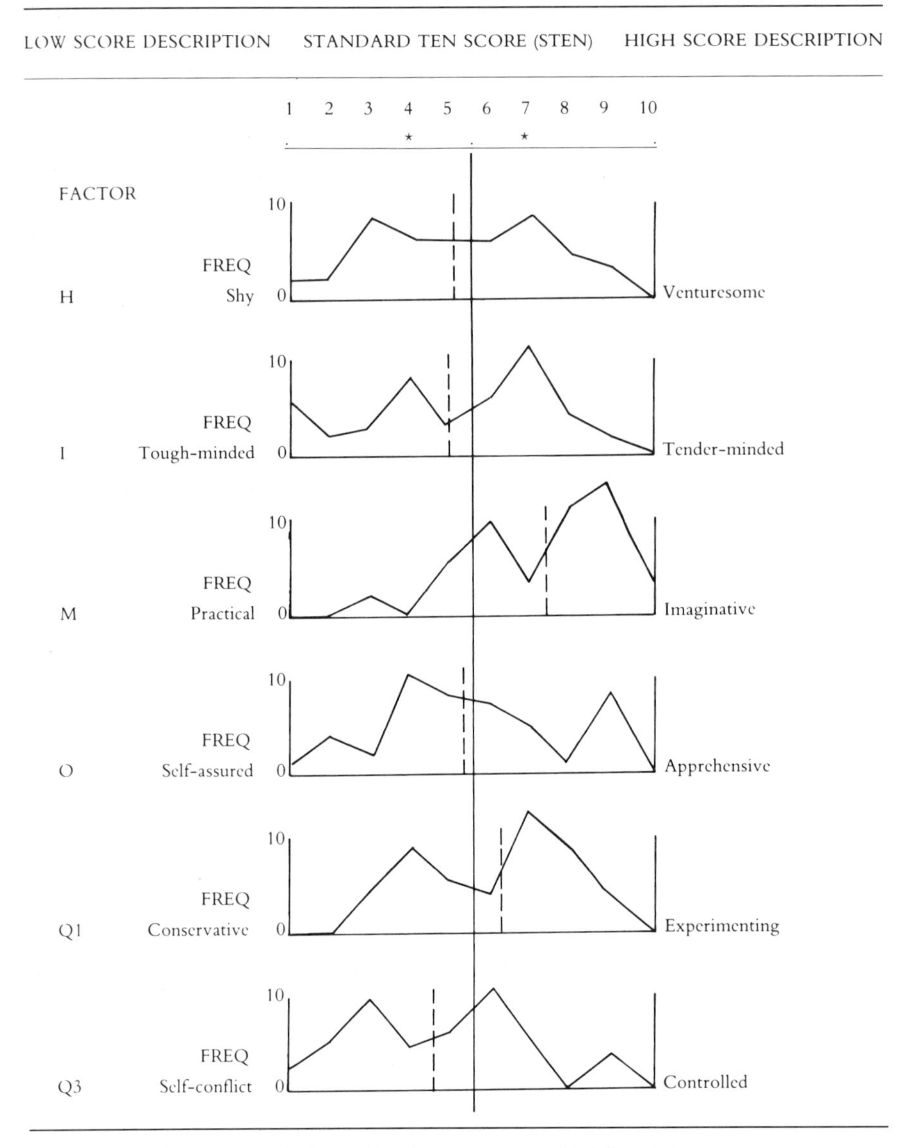

Figure 8.9 Dependents' 16PF results showing bi-modal distributions (asterisks define 1 S.D. from norm). Broken vertical line is group mean.

both modes lay above the mean for the norm.

The Dependents also showed consistently high scores for factor E (humble : assertive), with high scorers being described as assertive, aggressive, competitive and stubborn. Krug (1981) stated that such people were frequently forceful and very direct in their relationships, indicating a need to have their own way regardless of the feelings of others. These traits were described during the interviews, and were corroborated when spouses' opinions were elicited. The Dependents may have admitted to poor relationships with others but few would attempt to improve the situation by compromising their attitudes and opinions.

The Dependents' high scores for Q2 (group-dependent : self-sufficient), were also consistent with the personalities of the people interviewed. High scorers are typified by those who are accustomed to making their own decisions, who in group situations tend only to offer their own solutions rather than work as a team member, and are frequently rejected by others. They prefer to be alone and do not need or desire the support of others in order to function. While at school, high scorers tend to be successful, self-sufficient isolates, perhaps associating with just a few older friends. These results correspond closely to the Dependents' self-descriptions and once again high scores are found among scientists.

The Dependents also yielded results which were significantly higher than the general population for two of the Second Order factors, QIII and QIV. Factor QIII (tender minded emotionality : tough poise) is typified by those who show a tendency to think rather than to feel. Such thinkers tend to be enterprizing, alert and with a tendency to handle problems at a dry, cognitive, objective level. Their lack of feeling has been shown to lead them to miss the subtleties of life, but can also be seen as an effective way of inhibiting frustration and discouragement. The high levels which were seen for many of the group indicate individuals who become very detached and insensitive to the feelings of others.

High scores for factor QIV (subduedness : independence) are more frequently observed in males and indicate people who are aggressive, independent, incisive and a law unto themselves, and likely to exhibit considerable initiative. High scores have also been linked to Witkin's (1950) perceptual field independence, and both have been associated with high scores on factors B and E and low scores on factor A (Haynes and Carley, 1970). These trends were all observed in the Dependents' results from the 16PF and confirm the results from the GEFT.

Although all of these results were consistent with the results in earlier sections, one expected difference between the Dependents and the general adult population was not found. The early literature had suggested that the Dependents would tend to be shy, isolated and introverted, and this was apparently well supported by the results from the Leisure Activities Inventory, from the interviews, and in all the relevant comparisons made between the Dependents and the control groups. It had therefore been anticipated that the results from the 16PF would show the Dependents as a group yielding significantly lower scores on the Second Order factor QI, the introversion : extroversion scale, than the population norm. This was not found. The Dependents' results showed a normal distribution only very slightly lower than the population mean and well within the normal range. This discrepancy between the expected and observed result was considered at length.

If one assumes that the 16PF is able to measure introversion with some reliability one might conclude that the Dependents' were 'faking good' in their responses. However, as the 16PF was given shortly after the interviewees had described their own personalities, many of which included details of their poor and problematic social experiences, there was no reason to suppose that they would 'fake bad' during the interview but 'fake good' during the questionnaire. Furthermore, the Handbook for the 16PF states that the most successful method used to prevent distortion is to establish a good rapport with the testees. In all instances an effective rapport was established, often to the extent that personal information was disclosed of a depth which had not been anticipated. For this reason 'faking good' was not considered to provide an adequate explanation for the observed results. However, examination of the methods used to determine factor Q1 do go some way to explain why the Dependents' results did not show them to be introverted, as I and they would have expected.

Cattell *et al.*, (1970) considered that factor Q1 (introversion) was chiefly determined from five First Order factors (A, E, F, H and Q2) and when a more detailed examination was made of the 16PF results four of these, A, F, H and Q2, did yield results which suggested that the Dependents showed introverted tendencies. They were shown to be more reserved, sober and far more self-sufficient than the norms, and for factor H (shy : venturesome) there was a bimodal distribution. Only the result for factor E, which showed the Dependents as more assertive than humble, went against the expected results and this prevented the Dependents from yielding an overall result of introversion.

The Dependents were assertive and certainly not humble. they were very determined and independent, had great pride in their intellectual prowess and knowledge, and tended to consider themselves to be in many ways superior to the majority of the population. High scorers for factor E are described as assertive, aggressive, authoritative, competitive, hostile, and a law until themselves. These have been shown to be behaviours typical of some shy people (Litwinski, 1950; Kaplan, 1972), and could therefore surely be seen as the outward manifestations of people who are socially fearful or mistrustful, and not necessarily those of extroversion as believed by Cattell.

Similarities could also be drawn between the factors of assertiveness (factor E) and self-sufficiency (factor Q2), both of which were shown to be very strong traits of the Dependent group. Those who are self-sufficient are described as independent, resourceful, preferring to make their own decisions and having little need for feedback from others. This does not appear to contradict the descriptions of the assertive character but, in spite of any seeming similarities between them, assertiveness and self-sufficiency are negatively correlated for the purpose of measuring extroversion on the 16PF, and as they demand the same weighting their effects are thereby cancelled.

Others have also cast doubts on the factorial methods employed by Cattell in order to determine the Second Order factors for the 16PF. Warburton (1968), Saville (1978) and Ormerod and Billing (1982) all believed that factor E (humble : assertive) should be omitted when deriving the factor which would equate with Cattell's measure of extroversion.

Close scrutiny of the *Handbook for the 16PF* suggested that these two factors are not always negatively correlated. Indeed Cattell *et al.*, (1970) found assertiveness and self-sufficiency to be positively correlated when isolating pure factors from the source traits, and also when determining the Second Order factor QIV, independence. In addition, they were prepared to accept that in some instances certain groups of people could be classed as introverted in spite of the fact that their results on the Second Order factor QI (introversion : extroversion) appeared normal. This departure was used when the authors described the results from a group of 91 male physicists.

Although the physicists showed the expected low scores for factors A and F, and high scores for Q2, they also showed the 'extroverted' tendency for factor E, assertiveness, and for factor H, venturesome. These five results showed very similar trends as observed in the results from the Dependents. Although the Dependents were found to be more shy than the physicists, both yielded results for QI, introversion, which were within the normal range. Figure 8.10 shows the mean results from the Dependents and the physicists for the five major factors from which factor QI is determined. (The bias of the expected result for introversion is indicated by the underlining of the relevant polar term. The Dependents showed the expected bias on four occasions and the physicists only on three).

In spite of the physicists' normal results for factor QI, Cattell *et al.*, (1970) believed that this group were truly introverted, as expressed by their First Order factors, and they implied that the result for the Second Order factor QI was misleading. The authors went further to describe these results as illustrative of a 'special kind of introversion'. One can therefore conclude that in certain instances introversion can still be assumed to occur in spite of Second Order results to the contrary.

FOOTNOTE The personality profile for the physicists was the only one in the Handbook which bore any resemblance to that of the Dependents', with both groups exhibiting high scores for the Second Order factors of tough poise and independence, as well as those factors relating to introversion. However, there were too many differences to suggest that they showed similar personality factors on all levels. The physicists showed a lower level of anxiety and were more emotionally stable, trusting, practical, self-assured and controlled than the Dependents.

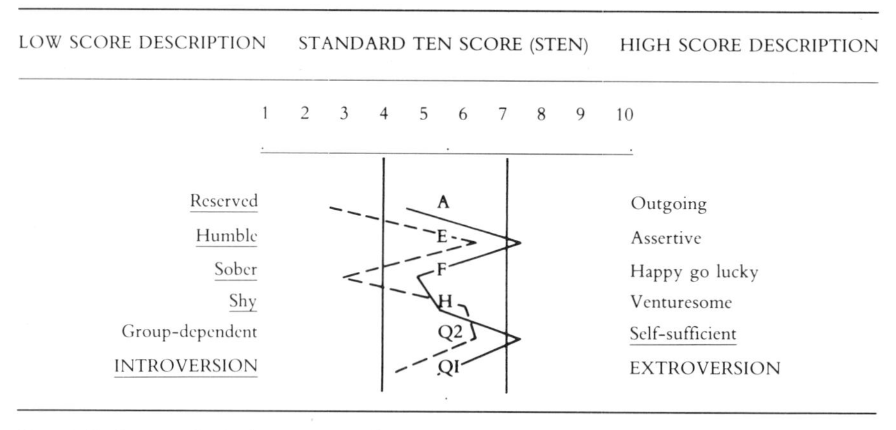

Figure 8.10 Sten scores for the five main primary factors used in the determination of factor QI, Extroversion : Introversion
Dependents N = 45 ______ , Physicists N = 91 _ _ _ _ (Cattell *et al*, 1970)

Because of the similarity of trends between the results from the physicists and the Dependents for these factors, the observed results from the 16PF therefore cannot be used to contradict all the other findings which suggested that the Dependents were introverted.

Discussion of personality factors

The results from the Self Report Scale suggested that in the main the self-ratings of the personalities of all three groups were very similar indeed. The Dependents were shown to be no more discontent, unfulfilled or unhappy than the controls, nor were they more intolerant, emotionally unstable or anxious. The three matched groups also showed similar levels of intelligence, ambition and impulsiveness, and the Dependents did not rate themselves as any more unconventional or non-conformist than the controls. Some significant differences were hightlighted by the use of the scales however, and as suggested by earlier results these tended to be confined to aspects of interpersonal relationships and social abilities. The Dependents described themselves as significantly more socially inhibited than the control groups.

Although many of the Dependents had during the interviews indicated that they had come to terms with their poor social skills and were uninterested in the opinions of others towards them, the use of scales to measure differences between Self and Ideal Self showed that they were not entirely happy with this aspect of their personalities. Ideally the Dependents would have wished for greater social ability in order to decrease their isolation from others, and the results showed that all three groups considered social skills to be desirable attributes.

Similarly, although all three groups wished to be more attractive than they considered themselves to be, the Dependents rated themselves as significantly less attractive than the control groups. Although not a personality trait, one wonders if the Dependents' view of themselves as unattractive was in direct response to the manner in which they felt they were treated by others, as the Self Report Scale also demonstrated that they felt misunderstood and isolated. Thomas and Young (1938) showed that people who were disliked were seen by others to be unattractive as well as introverted, and likewise Dion and Berscheid (1974) found that unpopular schoolboys were seen as unattractive and anti-social by their peers. It was therefore likely that the converse was true; that people who found it hard to make relationships considered themselves to be unattractive because of the reception they received from others.

Although unable to do more than make a subjective judgment, the interviewed Dependents appeared no more nor less attractive than average, ranging from the well-favoured to the plain, but as a group clearly felt themselves to be less attractive than average. Many did however give the impression that their dress and physical appearance were unimportant and irrelevant to them; a feature of the computer dependents which had been observed by Weizenbaum (1976), by Turkle (1984) who saw it as a typical example of male arrogance, and by Levy (1984) who felt it was representative of brilliant people 'who put things like physical appearance into their proper, trivial perspective.'

All three groups held similar characteristics to be ideal on the Ideal Self Scale, and perhaps such ratings are universally considered to be desirable and are not necessarily particular to these groups of people. There were nevertheless sufficient differences between the Dependents and the control groups on other measures to substantiate the interviewed Dependents' descriptions of themselves and to give a composite picture of the Dependent personality.

The Dependents reported that by school age many of them felt very isolated from their peers. Realization that their interests and hobbies were atypical of other children led many to feel themselves alienated and different from their fellows. Such feelings persisted, although finding some relief during secondary and tertiary education where a few like-minded companions were found. Relationships with the opposite sex did develop, although often with some difficulty, but even after marriage the Dependents' main areas of fulfilment lay with their work and hobbies rather than their families and friends.

The Dependents seemed from an early age to have been unable to share fully and intimately with others, and over the years their hobbies had provided their main source of satisfaction and fulfilment. However, interest in other activities had paled with the advent of the computer and their lives had become dominated by computing. A few interviewees freely admitted that their hobbies were used as substitutes or escapes from social activities, with the human-computer interaction seen to offer feelings of control, satisfaction and intellectual stimulation within a safe, non-threatening and confidence-building environment.

Certain of the results obtained, especially those which showed the Dependents to be both socially inhibited and field-independent, were consistent with other research findings about professional computer programmers. McClure and Mears (1984) and Kerr (1987) both suggested that introversion was a personality trait common to computer programmers, and none of the authors who offered descriptions of the computer dependent personality failed to include the factors of shyness or introversion with reference to them. Carlisle (1974) and Poulson (1983) also showed that successful programmers and those interested in computer hardware tended to have high abstract reasoning abilities and a field-independent cognitive style. The Dependents therefore exhibited some personality characteristics similar to those of computer programmers in general. This was perhaps of little surprise, even though not all of the Dependents programmed computers nor considered themselves to be efficient programmers. If software and hardware designers exhibit convergent, field-independent cognitive styles their products are more likely to be understood and easily used by those who share the same traits. As Runkel (1956) stated, good communication is facilitated when individuals share the same cognitive style and it was not therefore surprising that the Dependents found their interactions at the computer to be non-stressful. They were sharing the same patterns of communication as the software designers.

The discussion in Chapter 7 suggested that early parenting methods were significantly influential in the development of the object-centred personality, and these views were consistent with the results obtained from the Dependents' self-ratings and the results from the 16PF. The Dependents rated their mothers and fathers to be more distant than those of the controls and this interaction between parent and child, which frequently lacked outward signs of warmth and unconditional acceptance, seemed to have been highly instrumental in the development of the Dependents' subsequent interests, social lives and personalities.

Although causality cannot be proven, research has repeatedly shown that this type of parenting, especially towards boys, can lead to shyness, self-sufficiency, social inhibition and isolation, and towards a convergent, field-independent cognitive style which is objective, scientifically oriented and task-centred; all factors of the Dependents' personalities according to the results from the interviews, scales, the 16PF and the GEFT. Whether a person's activities are found to be towards objects or distanced from people, both are likely to be engendered by parenting of a somewhat rejecting, neglecting or distant type as experienced by the Dependents. The Dependents' need for solitary, intellectual pursuits, giving little regard to emotions, other people or everyday practicalities indicated that they had discovered and developed methods of living which were in keeping with their personalities and interests. One could liken the interests and the personalities of the Dependents to the spiritual detachment fundamental to many eastern philosophies. Such detachment from others, although a defensive move, can give serenity and security, and allow for the expression of creativity and original thinking. Weiss (1962) believed such behaviour to be used as a safe method of dealing with emotions and rejection, and Roe (1957), Hudson (1966) and Smail (1984) all looked upon the object-centred bias positively, seeing it as one used by many people as a very successful and effective means of dealing with life.

The results gave a clear picture of the personalities of the Dependents, who, from my own observation, did not appear on the whole to be any more unusual in their strategies for coping with life and relationships than many of the scientists and technologists within a typical university. Nevertheless, much of the literature describing the computer dependent personality did not view them as coming merely from the object-centred pole of the people- to object-centred continuum. More sinister and emotionally loaded descriptions were given, although none appears to have been based upon empirical, comparative research.

Isherwood (1987) labelled the computer dependent as suffering from a psychosis; Eysenck, a neurosis; Zimbardo (in Ingber, 1981), paranoia; and to Weizenbaum (1984) it represented 'a psychopathology that is far less ambiguous than, say, the milder forms of schizophrenia or paranoia . . . it *is* a compulsion'. Weizenbaum also described it as 'an abdication of responsibility, a closing of the mind to reality without an accompanying sense of incompleteness', and likened it to the Oedipal dilemma exhibited by the 'megalomaniacal fantasies of onmipotence'. Similarly, Thimbleby (1979) described it as a neurotic struggle to gain power over the father figure (the computer) or as a regression back to the protective environment of the womb (the computer room); and Brod (1984) as 'the hidden Freudian wish to uproot the father' by dominating the authority of new technology. Brod cited the specific case of a boy, unable to gain attention from his father, who turned to the computer. Although stating that this increased the child's isolation, he also realized that the boy was then able to get attention because of his computing abilities.

Although these authors have tended to imply that the effects of parenting methods were in some way associated with the computer dependent persons' subsequent behaviours, it was the computer which became the 'whipping boy', not the parents, and little empathy seemed to have been extended towards the Dependents and their need to turn to the computer for satisfaction and security. The authors' terminology suggested that computer dependency was a symptom of mental illness, not a coping strategy or a manifestation of a particular personality style, and that the computer was the cause of that 'illness'.

Such beliefs were not substantiated by the results from this research which indicated that the needs, personalities and object-centred biases of the Dependents had been stable from an early age, as was reflected by their preferred hobbies and work interests, and their social experiences. The activities enjoyed in earlier years contained many of the attributes found when using new technology, with computing being a natural progression from them. Computing had not appeared to alter their behaviours and needs, although it had been able to satisfy them more fully; more hours could be devoted to this activity than to many others, and it could more easily be integrated into their working lives than most of their other past interests. There was no suggestion from the results that people who had been gregarious and extroverted had, since using the computer, become more object-centred in their outlooks or behaviours.

Only a very small proportion of the Dependents described extreme and overt behaviours during their lives which could be labelled as neurotic, but even the use of this term is open to various interpretations. Eysenck (1977) described neurotic behaviour as that which is maladaptive, seen by the person to be irrational, 'absurd or irrelevant, but which he is powerless to change'. This may have been an attitude held by the spouses of a few Dependents but was not one which applied to the Dependents themselves. Most had no desire at all to alter their behaviour but merely wished to change the attitudes of others towards them. Conversely, Eysenck described normal behaviour as that occurring when the person, knowing what would satisfy him 'and, given the choice, he knows what he wants and grabs it'. The second description far more accurately described the behaviour of the Dependents, who were able to express their true personalities by spending most of their waking hours in intellectually stimulating interaction with a computer.

Definitions vary enormously, and Eysenck's description of 'normal' bore strong resemblance to that which Fromm (1941) considered to be neurotic.

> 'The person who is normal in terms of being well adapted is often less healthy than the neurotic person in terms of human values. Often he is well adapted only at the expense of having given up his self in order to become more or less the person he believes he is expected to be. All genuine individuality and spontaneity may have been lost . . . (the neurotic) is less crippled than the kind of normal person who has lost his individuality altogether.'

Confirming Fromm's observation, many of the Dependents seemed to have refused to comply with what was expected by society; they did not find social activities enjoyable, failed to understand what was expected of them in such situations, and instead of submitting themselves to unhappiness and distress had followed their own routes to contentment.

The range of coping strategies employed by the Dependents, suggested to be positively oriented towards objects, were different from but perhaps no more extreme than would have been found when interviewing any other group of individuals. Culpin (1948) estimated that, from a study of 1 000 people, about 60 per cent showed some signs of 'neurotic symptoms', and Valentine (1956) in his book '*The normal child and some of his abnormalities*', as suggested by the title of his book, also adhered to the belief that few of us are 'normal' in all our methods of dealing with the world. The latter described the isolated, intelligent child, immersed in his hobbies, as one who copes in the world by the use of untypical strategies. Such children have always been with us and are usually capable of successfully managing their lives through the solitary pursuit of intellectual stimulation.

However, in spite of the fact that the majority of the Dependents seemed to have found extremely successful ways of dealing with the world, there was little doubt that a few of the Dependents had had more to contend with than most. Some had indeed experienced great trauma or unhappiness in their earlier years, and from the interviews it was not difficult to appreciate why they had turned to artefacts for the satisfaction of their needs.

The schizoid personality type

Although the Dependents did not exhibit the extreme mental states suggested by other authors, this investigation demonstrated that many of them did exhibit tendencies which could be described as typical of the 'schizoid' personality type. (Schizoid is a term used to describe a personality type not a mental illness. Unfortunately it is frequently misunderstood by the layman who often confuses it with schizophrenia.) The *Penguin Dictionary of Psychology* (Drever, 1952) defines schizoid as 'a *personality type*, tending towards *dissociation* of the emotional from the intellectual life'.

Schizoid traits and personalities are very common (Cull *et al.*, 1984), and are reflected in those people 'whose general stability in any reasonable environment is quite adequate, but who clearly lack the capacity for simple, spontaneous, warm and friendly responsiveness to their human kind' (Guntrip, 1968); a personality type which he states 'cannot be set apart from the so-called "normal"'. Similarly, MacKay (1975) believes such a person to be in contact with reality but 'prefers to live in his own world of day-dreams and fantasies and avoids close personal relationships'. Both Turkle (1984) and Gray (1984) used this term when describing the computer dependent personality, with the latter concluding that it was a type frequently observed within the scientific community.

It is a personality type more common in males than females (Chick *et al.*, 1979), and its main feature is described as a 'defect in the capacity to form social relationships' not due to mental or psychotic disorder (in the *Diagnostic and Statistical Manual of Mental Disorders, DSM III;* American Psychiatric Association, 1980). In childhood this is characterized by having no close friends of a similar age other than a relative or a similarly socially isolated child; no apparent interest in making friends; no pleasure from, and general avoidance of, peer interactions; and no interest in activities that involve other children. In adulthood, individuals 'show little or no desire for social involvement, usually preferring to be "loners", and have few, if any, close friends. They appear reserved, withdrawn, and reclusive and usually pursue solitary interests or hobbies.' These descriptions bore very close relationship to those offered by the Dependent interviewees. *DSM III* also states that such a personality type is no hindrance to high achievement in a working environment where social interaction is unnecessary, although problems can occur when social demands are made. This was consistent with the Dependents' descriptions of their work experiences, where some chose to avoid managerial responsibility because of their poor interaction with others.

Lake (1966) described children showing schizoid tendencies to be intelligent, serious and 'old before their years', believing them to turn to a 'world of detached mental activity' as compensation for the lack of satisfaction found with the real world. Included among Lake's lengthy descriptions of traits and attitudes attributed to the schizoid personality were many which directly applied to the Dependents, as follows:

- Introversion;
- Defence by detachment from persons;
- Time spent peacefully in solitariness;
- Awkward on social occasions, unless role-playing;
- Recovers well-being in solitude;
- Eccentric, atypical, not gregarious; or in a clique of eccentrics or outsiders;
- Militantly self-sufficient;
- Action negatively focused upon exterior goals, positively focused upon interior goals;
- Pride in mind and knowledge;
- Pursuit of perennial wisdom and abstractions;
- If highly intelligent utilizes defence by intellectualism;
- Refuses to be tied down to the necessities of existence.

These patterns of behaviour, which may be described as defensive, are established in order to maintain a state of equilibrium when dealing with people and life, and if successful these strategies become relatively stable and remain the dominant means of coping. The theories surrounding the development of the schizoid personality all stem back to early childhood and parenting methods.

Incorporating the theories of Freud, Klein, Fairburn and Winnicott, Guntrip (1968) put forward a very comprehensive thesis concerning the schizoid personality. He believed it develops because of the lack of security felt by a child when love is conditional or withdrawn periodically, as is found when behaviour and achievement are rewarded rather than the person. The child's needs are then not fully met and the parental behaviour is interpreted as a cessation of love, and rejection. 'The schizoid person's capacity to love has been frozen by early experiences of rejection and the breakdown of real life relationships' (Guntrip, 1968). He saw it as a post-natal development, apparent in the early years, not as an hereditary factor nor one which would occur for the first time in adulthood.

Similarly, Horney (1946, 1951) believed that this type of upbringing, instead of leading to the development of confidence in self and others, created a basic anxiety. In order to cope with such anxiety she suggested that many children develop the need to move away from people, and although seeing this as a disturbance of human relationships she recognized it as necessary for the purpose of self-integration. This distancing from others then develops into a desire for control, perfection, self-sufficiency, privacy and independence, which in turn may lead to feelings of superiority and uniqueness from others. By controlling the emotions through detachment and aloofness, Horney believed that more emphasis becomes placed upon intelligence and the intellectual side of life.

This was confirmed by Smail (1984) who, with reference to the schizoid personality type, stated that 'it is from one's parents that one first gains an idea of whether one is lovable'. In discussion of the damaging effects of love conditional upon performance, Smail suggests that middle-class families in particular use the manipulation of love and affection in their dealings with their children, which lead to the sapping of confidence in the child's self-esteem and self-worth. Without the ability to alter or control the external reality of this type of parenting, an internal world is created by the child which can be controlled and manipulated. From this stance, the control of the environment and its artefacts develops and becomes dominant.

The personalities and interests of the Dependents were consistent with the theories of Roe (1957) and Hudson (1966), which the former had also causally linked to rejecting or neglecting parenting. Each person adopts a strategy of coping as an effective mechanism for dealing with life, and although such mechanisms may differ according to one's circumstances neither Roe nor Hudson considered the object-centred reaction to be neurotic in character.

> 'Whatever the moral, it is not that some occupational groups are more neurotic than others. Although convergers and divergers use different tactics in dealing with the pressures of work and emotional experience,

> one tactic is not necessarily better or worse than the other. Each has its characteristic strengths and weaknesses; and the neurotic is not the man who adopts a particular intellectual and personal style, but the one who, having adopted a style, suffers its weaknesses without enjoying its strengths.'
>
> Hudson (1966)

The Dependent personality and the computer

The self-reports and the results from the tests suggested that the personality profiles of the group of Dependents did conform to the descriptions of the schizoid personality. Throughout all generations there have no doubt been object-centred, shy people who have turned away from human relationships, have subjugated their emotions, and have resorted to solitary activities in order to find satisfaction. However, never before has there been an activity such as computing which could give the distinct impression of providing companionship and partnership, to which even the keeping of devoted and non-judgmental pets cannot compare. Pets can supply the emotional warmth and acceptance, but provide little in the way of intellectual stimulation to those who demand it. Satisfaction of intellectual challenges was consistently found to be one of the Dependents' primary drives, and it is perhaps understandable that they have appeared to find satisfaction when dealing with a logical, 'intelligent' machine. The schizoid personality type's need for an escape into fantasy is well catered for, and if one is adept with new technology it can provide an endless supply of positive feedback.

Some might deduce that the feedback from a computer is too similar to that of the parent who loves conditionally to warrant the attention that the Dependents devote to it; the exploratory use of a computer could appear masochistic in the extreme to the hostile or suspicious user. Computing demands strict adherence to rules, is intolerant of small errors, appears temperamental, is unemotional, and appears non-forgiving. However, such descriptions of the computer would either appear alien to the object-centred, technologically-oriented Dependent, or would be classed as the positive features of their interaction with the computer.

The Dependents were not fearful of the machine nor suspicious of its powers, they were excited about their opportunities to use new technology, and from their studies, including those in science fiction, had some knowledge of how they functioned. They understood its rules, which to them seem logical and rational unlike those expressed by humans, and the interaction offered a secure base from which to develop their abilities and interests. Although in some ways the computer may act as a parental figure, by rewarding good behaviour (the program will run), it does not appear to punish the Dependent user by a withdrawal of 'love' (when the program does not run) but waits patiently for the next input. The feedback from the computer tends merely to ignore poor performance, does not judge the whole person as inadequate, but allows one to try again without permanent damage to the ego.

Like parents who reward achievement and have great expectations for their offspring, so too the computer is never completely satisfied; there are always more goals to strive for. However, for the Dependents these goals are totally under their command and control, are realistic, and are not superimposed from without. They are able to develop their programs as far as they wish; they are the only judge of their accomplishments as the computer has no biases and remains impartial. Programming allows them to work at their own pace, make mistakes, experiment with new ideas, learn new techniques, and develop their personal interests to the full. In addition, the computer was not always viewed as totally without emotion, although the Dependents were well aware of its inability to do more than compute in simple arithmetic terms. It was sometimes viewed as a true friend, one which would not withdraw its favours if a mistake was made, it would not switch itself off, but was an ever-faithful, encouraging helpmate.

> 'With computers you got justice. What you put in you got out. Instant results . . . If you worked hard enough, *you understood* . . . not like life. There were rewards for accomplishment . . . not like normal life.'
>
> Bishoff (1983)

The interviews with the Dependents had indicated that most had experienced great difficulty meeting like-minded friends with whom to share their ideas; however, the computer had been able to act as the perfect substitute. It reacted in an intelligent manner, reflected the Dependents' interests, offered reinforcing intermittent rewards, and satisfied their internal goals and needs. It offered them purpose, a medium for the achievement of success and accomplishment, and an outlet for fantasy and escapism. Their needs were gratified; it reduced anxiety and boosted their self-esteem.

Instead of experiencing frustration and stress while attempting to cope in a world geared to successful social interaction, they were able to relax and via the keyboard were able to achieve the type of stimulating encounter for which they had long been searching. The preferred leisure activities of the Dependents showed a history of hobbies which were undertaken for their own intrinsic worth; they were on the whole not put to practical use but provided intellectual stimulation and the ability to understand and control their environment. The Dependents showed little need for others, tended to introversion, and were highly independent and self-sufficient. They did not rate themselves as more unfulfilled, unhappy or anxious than the controls, and the interviews showed that most had reached a state of equilibrium within their lives which did offer fulfilment and happiness. This was in part due to the stabilizing, caring influence of the partners of those who had married; but for the majority the advent of the computer had provided them with the perfect medium to utilize their abilities and aptitudes as no other person or artefact had been able to do before.

From such results one might surmise that all who chose to devote themselves almost exclusively to the pursuit of intellectual stimulation and the control of artefacts would be found to exhibit a schizoid personality type, show an object-centred, convergent bias, and have a history of shyness and poor social relationships, as exhibited by the Dependents. Although this cannot be established from the results of this research, suffice to say that these factors differentiated the Dependents from the control groups and from the population norms with whom they were compared.

The need to have control over aspects of one's environment seems especially necessary for those who have not been loved unconditionally by their parents. This tendency was shown to be quite common within society, and from the interviews I would suggest that the need to control the computer is neither neurotic nor pathological but provides an admirable means of coping for those who may previously have felt inadequately fulfilled.

FOOTNOTE Appendix 1 contains a number of case studies drawn from those who were interviewed. These are offered as illustrations of the backgrounds, lifestyles, needs, and computing activities undertaken by those specific individuals. These case studies were not selected to represent the 'typical' or 'average' computer dependent person, as such a person did not exist, but the results from the comparative data were able to illustrate how the Dependents differed in general from the control groups.

The case studies were chosen to illustrate the most extreme problems experienced by some individuals and to reflect the most overt characteristics exhibited by the group. From these one may more easily deduce why intellectual activities and the study of inanimate artefacts had come to dominate the lives of the Dependents and act as coping strategies.

(The labels attached to each case study, either 'Networker', 'Worker' or 'Explorer', refer to the three different types of computer dependent personality which were isolated. These are described and discussed in the following chapter.)

Chapter 9
Different types of computer dependency

Introduction

Although there were many similarities between the Dependent respondents, and numerous differences between them and the control groups, it became apparent from the analyses already undertaken that the Dependents did not form an homogeneous group, in spite of the fact that they all considered themselves to be dependent upon computers. The interviews highlighted differences in personality, and in the use of and attitudes held towards the computer by the individual Dependents. Because of this lack of cohesiveness, differences were looked for amongst the Dependents to observe whether distinct sub-groups were apparent.

Searching for sub-groups

Two of the most obvious variables upon which to divide the group were by age and by sex. However, initial passes through the data indicated that gross differences were not apparent using these divisions. There were too few females to make any analyses reliable between the sexes, and the variety of attributes seen in the young were invariably observed in the older members. Two further divisions were examined in more depth and the interviewees were divided (a) according to the time spent on their home computers and (b) according to their preferred computing activities.

If the severity of dependency upon computers could be measured by the amount of spare time spent computing, one would expect there to be measurable differences for many other variables between those who used the computer for a short time per week and those who used it extensively. In order to examine this hypothesis the interviewees were subdivided according to whether their home computing time lay above or below the mean time observed of 25·7 hours per week.

In addition, the interviewees appeared to have developed their own individual areas of specialization and appeared to differ in personality and needs according to the main application and use of their home computer. For example, those who used the computer chiefly as a means of making contact with others appeared also to report fewer social difficulties than those whose activities remained solitary. A question had been included in the early General Questionnaire which asked all of the respondents to indicate to which of four main computing activities they preferred to devote the majority of their time, as follows:

(a) Exploring the system and self-education;
(b) Writing specific programs for an end-product;
(c) Using commercial software; and
(d) Networking.

The interviewees revealed that although many were versatile in their use of computers and did practise a number of different types of activities, one particular activity was always predominant. None of the interviewees had spent less than two years dependent upon computers when interviewed and all seemed settled into their preferred activity, tending only to vary their tasks for the sake of variety and to keep up to date with new hardware and software developments.

The results from the interviews showed that there were in fact not four distinct groups amongst the Dependent interviewees but three, as those who programmed with a definite end-product in mind were also those who frequently used commercial software. The 45 interviewees were subdivided as follows, with each group being given a descriptive title:

(1) A group of 12 *Networkers* who rarely, if ever, wrote their own programs. They used the computer as a means of communication via the networks with other databases and individuals, to gain computing information, to engage in social interaction, to play MUD (an interactive computer game), or for hacking.

(2) A group of eight *Workers*, all of whom stated that their computing was centred upon work-related activities consisting of structured programming and the use of commercial software.

(3) The largest group comprised 25 *Explorers*, who stated that they spent the majority of their time programming in an investigative, self-educational and exploratory manner.

Differences between the sub-groups

Divisions according to time spent and preferred computing activities

The significant results found when the 109 questionnaire items were subdivided and re-analyzed are shown in Table 9.1. The first column

('Time Spent') lists the significant results obtained when the group was divided according to whether their time spent computing at home lay above or below the observed mean time, and the second column ('Activity') lists those obtained when the group was divided according to their preferred computing activities. Only eight of the 109 results analyzed showed significant differences when the interviewees were divided according to the amount of their spare time devoted to computing, therefore this avenue was not pursued further. However, when the Dependents were divided according to their preferred computing activities (*i.e.*, into Networkers, Workers and Explorers), 43 of the 109 analyses yielded significant results. The remaining data were then divided according to the Dependents' preferred computing activities to examine the full extent of the differences between these three groups.

Not only were differences observed between the Networkers, the Workers and the Explorers in various aspects of their computing activities and attitudes towards computers, but also in their schooling, family relationships and social behaviours. As had been observed during the interviews, these results suggested that the Dependents' chosen computing activities were indeed related to behaviours and attitudes found in other areas of life.

Comparisons between the Networkers, Workers and Explorers

Content analysis from the results of open-ended questions showed that the Networkers, Workers and Explorers held different attitudes towards computers and computing and also attributed different reasons to their computer dependency. These significant differences between the groups are summarized in Table 9.2.

Although there was overlap between the discriptions from the groups, the results indicated that significantly more of the Networkers tended to view the computer as a toy, a tool used for fun and entertainment; the Workers treated computing with more seriousness and saw it as a tool used to increase competence; while the Explorers tended to anthropomorphize the computer, using it both as entertainment and also as an escape from social relationships.

Eleven of the 20 First and Second Order factors of the 16PF also showed significant differences between the Networkers, the Workers and the Explorers. The profiles of these significant results are shown in Figure 9.1. The results are illustrated by the mean scores of each group, and where the results from two groups were not significantly different their combined mean scores are shown.

The results from this reanalysis of the 16PF data showed the Networkers to be more assertive, venturesome, imaginative, extrovert and independent than the other two groups; the Workers to be more

Table 9.1 Differences obtained from questionnaire data when the Dependents were divided by time spent computing and type of activity preferred

Questionnaire items	Time spent	Activity
Enjoy primary school	–	**
Enjoy secondary school	–	*
Type of secondary school	–	*
Sporty at school	–	*
Take on responsibilities at school	–	*
Type of books read as a child	–	*
Solitary as a child	–	**
Shy as a child	–	*
Find it easy to make friends now	–	*
Feel shy now	–	***
Favourite of one parent	–	*
Identify with parent of same sex as self	–	**
How chose to spend free evening	–	*
Who live with now	–	*
Job makes use of talents	–	**
Job controlled by others	–	***
Differentiate between work and play	–	*
More important, home or work computing	–	**
Self-employed	–	***
Number of computers owned	–	**
Value of hardware	–	**
Need to understand the hardware	–	*
Preferred computing activity	–	***
Time spent computing at home (hrs/wk)	***	–
Time spent computing at work (hrs/wk)	–	*
Time spent computing in total (hrs/wk)	*	*
Number of days spent computing	*	–
Percentage of computing time spent programming	–	*
Percentage programming time spent exploring	–	*
Partners' rating of spare time spent computing	–	*
Self-rating of dependency, scale 0–100	*	–
Find it difficult to stop when computing	–	*
BASIC only programming language used	*	**
Preferred programming language	–	*
Use of machine code	–	**
Self-taught or trained in programming	*	–
Preplanned or hands-on programming	–	***
Use of graphics/sound in programs	–	*
Useful end-products from computing	–	*
Programs considered well-written	–	*
Feel creative when computing	–	*
Computing has broadened interests	**	–
Computing has broadened intellect	–	*
Use of the networks	–	**
Preferred networking activity	–	*
Software bought or copied	–	**
Enjoy computer games	*	**
Attitude towards hacking	–	***

Table 9.2 Significant results from open-ended descriptions of computers and computing, between Networkers, Workers and Explorers

View of computers/computing	Networkers	Workers	Explorers	p
Networkers different from other two groups				
Computer seen as toy/hobby	Yes	No	No	***
Workers different from other two groups				
Computing is intellectual challenge	Yes	No	Yes	*
Computing is fun/entertainment	Yes	No	Yes	**
Computing increases competence	No	Yes	No	*
Explorers different from other two groups				
Computer seen as tool/machine	Yes	Yes	No	**
Computer seen as friend/companion	No	No	Yes	*
Computer used as compensation/escape	No	No	Yes	*
Computer easier than people	No	No	Yes	*

conscientious, controlled and relaxed; and the Explorers to be more affected by feelings and more apprehensive, tense and anxious.

All of the remaining results (from the questionnaires, the GEFT, the six scales and the Leisure Activities Inventory) which showed significant differences between the Networkers, the Workers and the Explorers are summarized in Tables 9.3 to 9.5 in the following sections.

Discussion

This discussion is divided into four sections according to the results which showed:

- No significant differences between the groups;
- The Networkers different from the Workers and Explorers;
- The Workers different from the Networkers and Explorers;
- The Explorers different from the Networkers and Workers.

No significant differences between the groups

There were no significant differences between the three groups for the demographic variables analyzed. Their ages, sex distribution, marital status, educational levels, favourite subject areas and birth order showed that the three groups were drawn from the same population. Nor were there differences for the majority of questions which investigated intellectual interests, their family lives and partnerships. Such variables were not therefore influential in the choice of the groups' individual preference in computer activities.

The majority of all three groups considered themselves to have felt 'different' from and more serious than other children, and as adults no

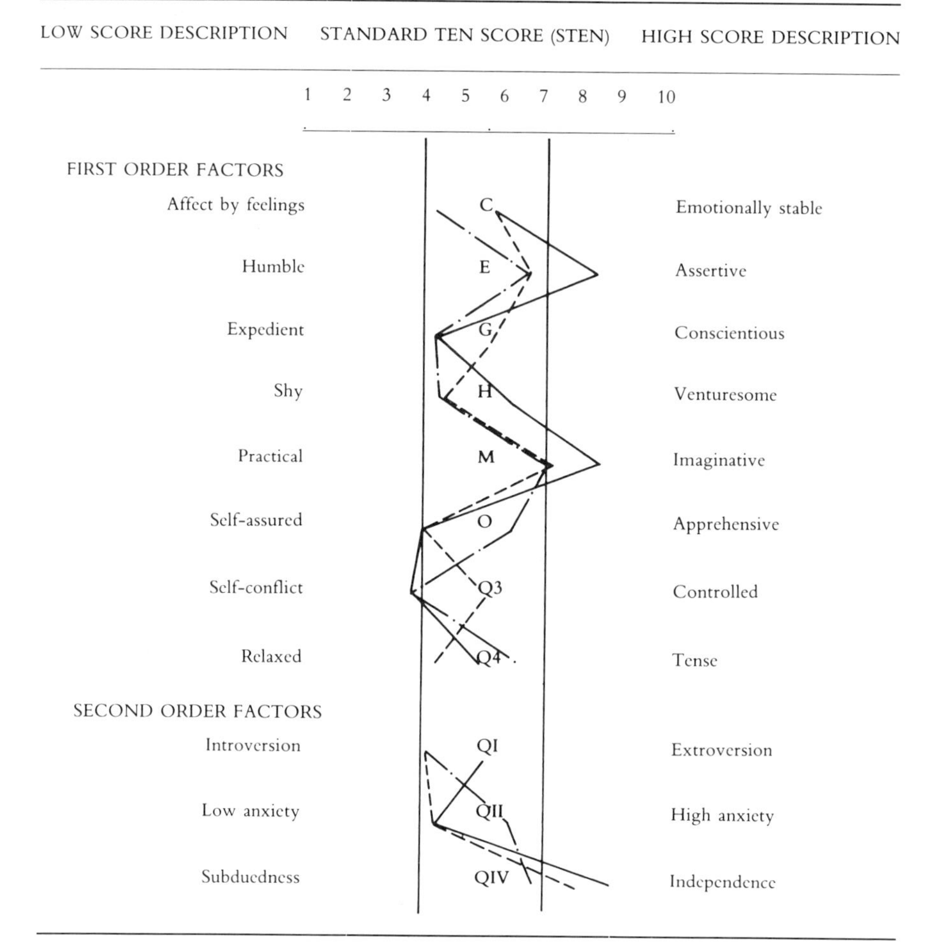

Figure 9.1 Means of 16PF sten scores showing significant differences between the Networkers, Workers and Explorers
——— Networkers, - - - - *Workers*, .—.—. *Explorers*

one group spent more evenings socializing than any other. All three were significantly more interested in science and technology than the arts, and as adults their preferred reading matter centred upon non-fiction and science fiction. Most had developed a way of life which they found fulfilling, and felt content with their lives. Even their employment was seen more as play than work, as was confirmed by the fact that the majority of all three groups stated that they worked for enjoyment rather than necessity.

There were no significant differences between the groups in the estimated times they spent computing at home, the number of days they used their home computers, nor in the proportion of their spare time they felt that they devoted to computing. All three groups had been dependent for similar numbers of years, and both they and their partners gave similar assessments for their levels of dependency. No one group felt that their computing caused more problems to themselves or others, nor did they differ in their own or others' attempts to ration their computing.

Most had learnt to program before owning a computer and considered themselves to be self-taught. All three groups owned similar quantities of software which they had often adapted, and similar numbers from each group had attempted hacking. As would be expected from people who had volunteered to take part in this research, the majority of all three groups also stated that computing was their best hobby and gave them their main fulfilment in life.

The set of non-significant results confirmed to a great extent that there were fundamental similarities between the groups. They were drawn from a similar population, and each group appeared to be as equally dependent upon computers as the other, devoting similar proportions of their spare time to their activities. Perhaps, more importantly, no one type of preferred activity was shown to be more or less likely to cause problems to the individuals or to others than any other (the effects of computer dependency are discussed in the following chapter). The differences which were observed between the three groups, however, did indicate that they differed psychologically, and that their choice of computing activity was congruent with their personalities.

The Networkers

The results which differentiated the Networkers from the other two groups, the Workers and the Explorers, are shown in Table 9.3, and these are clarified and expanded with data from the interviews. The overriding difference between the Networkers and the other interviewees lay in their more positive attitudes towards people and social activities. They considered themselves to be less solitary and shy as children and adults, and found it easier to make friends. Their attitudes towards people confirmed that they viewed social interactions as less frightening and more relaxing than did the other groups. The Networkers were more likely to have enjoyed their secondary schooling perhaps because of these factors, and were also more interested in entertaining and holidays than the other Dependents, activities which usually involve interaction with others. Although more likely to have attended discotheques, they were no more interested in this type of entertainment than the other groups, and like them did not enjoy parties and other large, unstructured events, preferring small, intimate gatherings with people of similar interests.

The majority of the Networkers, unlike most of the other Dependents, appeared also to have experienced far fewer difficulties in their

Table 9.3 Results differentiating the Networkers from other Dependents

Results which differentiated the Networkers from others			p
Questionnaire and interview data			
Less likely to have been solitary as a child			★★
Less likely to have been shy as a child			★
More likely to have enjoyed secondary school			★
Less likely to report themselves as being shy now			★★★
More likely to find it easy to make friends now			★
More likely to have had numerous partners			★★
More likely to identify with parent of the same sex as self			★★
Parents less likely to have had favourites			★
Mothers more likely to be in paid employment			★
Preferred activity: time spent networking			★★★
More likely to own a modem			★★
Spent less hours computing in total (mean 35·5 hrs/week)			★
Spent a smaller percentage of computer time programming			★★★
More likely to be only familiar with Basic language			★★
More likely to select Basic as preferred programming language			★
Less likely to be familiar with machine code			★★
Less use of graphics/sound when programming			★
Leisure activities inventory			
Networkers	more attempts at	Discotheques	★
	less attempts at	Radio-controlled models	★
		Electronics	★
		Weight training	★
Networkers	more interested in	Computer networking	★★
		Entertaining	★
		Holidays	★
Networkers	less interested in	Computer graphics	★★
		Computer programming	★
		Jigsaw puzzles	★
		Maths/logic puzzles	★
		Running	★
16PF Questionnaire			
Factor E	Humble : Assertive	Networkers more assertive	★
Factor H	Shy : Venturesome	more venturesome	★
Factor M	Practical. : Imaginative	more imaginative	★
Factor QI	Introversion : Extroversion	more extroverted	★
Factor QIV	Subduedness : Independence	more independent	★★
Self Report Scale			
Networkers	saw themselves as	less anxious	★
		more confident	★★
		more extroverted	★
		more imaginative	★
		more optimistic	★
		more happy	★
		more out-going	★
Attitudes towards computers			
Networkers	saw computers as	less satisfying	★
Attitude towards people			
Networkers	saw people as	less frightening	★
		more relaxing	★

relationships with the opposite sex. Most stated during the interviews that they had had numerous partners, while the majority of the Workers and the Explorers had often only had one or two partners, if any at all. Although finding it relatively easy to make friends with the opposite sex, they found interactions with people of their own sex to be virtually as difficult as did the rest of the Dependents. The men in particular seemed not to hold other males in very high esteem unless they shared mutual interests, preferring to confine their social activities within the family and marriage.

In spite of these measurable differences between the Networkers and the other interviewees, they were not free from social difficulties but were merely less socially inhibited than the Workers and the Explorers. The Networkers tended only to have few very close friends with whom they could share intimacies, and still felt themselves to be more shy than average, and not gregarious. Although the results seem to suggest that the Networkers had a more positive attitude towards life than the other groups, they still invariably demonstrated a schizoid personality type, and some still used the computer as an escape from the perceived mundaneness of life (see especially case study B, Appendix 1).

Although few measurable results yielded significant differences between their family backgrounds and those of the other groups, the Networkers appeared to have better relationships with their parents and more often than not during the interviews described their upbringing as happy. Although there was wide variation in their description of their parents, unlike the other interviewees the Networkers tended to concentrate upon the favourable aspects of their parents' personalities. Their parents were often described as encouraging and loving, although still not overtly so, with fathers tending to be seen as quiet, logical, creative, wise and hard-working and mothers as caring, strong, understanding and straightforward. Two women were the only ones within this group who offered negative opinions about their mothers, with one being described as strict and cold, the other as argumentative and disorganized. The male interviewees, in comparison with those from the other groups, were significantly more likely to identify with their fathers, and to speak more positively about them. In addition, the Networkers stated that there was less likelihood of one child within the family being treated as the favourite, and this lack of partiality could have been responsible for providing greater emotional and social stability than observed in the families of some of the other Dependents.

The Networkers' ratings of their own personalities on the Self Scale were significantly more positive than those of the Workers and Explorers. Their results showed them to be less anxious and more confident, optimistic, happy and out-going than the others. Their descriptions of being less shy and more extrovert were confirmed by the differences shown between the groups for the 16PF results, with their higher rating for extroversion being caused by their raised scores for factor H (shy : venturesome) and factor E (humble : assertive). Although the whole group of Dependents were shown on the 16PF to be independent and imaginative, the Networkers were found to show higher scores for these two factors than the Workers and the Explorers. The Networkers did not find it as difficult as the others to form and maintain relationships, and their raised scores on the 16PF suggested that they were also self-assured, self-motivated, unconventional, imaginatively creative, uninhibited and incisive people (IPAT, 1979). People found to display this 16PF profile are also, according to Krug (1981), shown to have a strong need for autonomy and control, and to exhibit a certain amount of emotional detachment.

Further results showed that the Networkers viewed and used their computers in ways different from the other Dependents. Although spending as much of their spare time on their home computers as the others, this time was far less likely to be spent programming. They knew fewer programming languages, were not interested in computer graphics and were less likely to have come into computing from a prior interest in electronics. Neither were they as interested as the other groups in puzzles, perhaps a necessary precursor for an interest in programming.

This group was more likely than the others to view the computer merely as a toy and a hobby, used for fun and entertainment (see Table 9.2) but was not in itself found to be of great intrinsic interest. For this reason they tended to find computers less satisfying than the Workers or the Explorers; perhaps this is to be expected from a group for whom the computer was used as a means to an end, not an end in itself, and as a vehicle for other activities. The computer was used as a communication device and not primarily for its computational powers; it could be seen as a sophisticated enhancement to the telephone system more than anything else.

This was the only group whose computing was not a lonely, isolated activity, but one in which they made contact with other individuals. However, although all of the Networkers spent time making contact with other databases, some as far away as the USA, in general communication with individuals or using the bulletin boards and private 'mailboxes', such activities were not of prime importance to most of them. The interviews revealed that games-playing and hacking were of the greatest interest, and most of their personal communication with others centred around these activities.

Some of the Networkers devoted much of their time to playing adventure games, especially MUD (Multi-user Dungeons and Dragons) a very elaborate, interactive game, at that time played via the Essex University DEC-10 mainframe computer. MUD is a text-based game of fantastic scenarios through which the players have to find their way in pursuit of treasure, solving puzzles along the way. Players include wizards, mortals, goblins and sorcerers, using swords and magic wands to defeat rivals in the pursuit of their goals. (See articles about MUD by Bartles, one of the game's devisors, published regularly in *Micro Adventurer* during 1984/5, and by Machin, 1984).

Two or three dozen people, with access to this computer via their modems, could play MUD simultaneously. Unfortunately for the players, MUD could only be accessed during the early hours of the morning when the university's computer was at rest from its more academic functions (from 1 am to 7 am on weekdays, and from 10 pm to 7 am at weekends). However, this did not deter some devotees from playing nightly. Therefore in order to play MUD one not only had to have a modem and suitable communications software, but had also to forfeit sleep, cope with the problems arising within the family because of the hours spent playing, and be prepared for large telephone bills. The latter problem could be eliminated by the strategic use of a little hacking, however. Some who were au fait with MUD managed to ensure that hours of games-playing were not attributed to their own telephone accounts, as various safe numbers were circulated amongst the players to ensure that others bore the brunt of the expense. Two people who acted as part-time advisors for a national computer network had their telephone bills paid for them, which eliminated this problem while also keeping them on the right side of the law.

There was little doubt that playing MUD was exciting and stimulating. After one long evening interview, wishing to experience the game first hand, I agreed to join a player as he prepared to access Essex. Not all who wished to play could, as it was strictly on a 'first come, first served' basis, and as the methods of access were not straightforward much of the excitement seemed to hinge upon whether one could gain entry. While waiting to see if his efforts had been successful, the interviewee thrust his wrist to me to feel his racing pulse. He did not get in, but stated that he always got an 'adrenalin high' before and during play. After much concentrated play, he had reached the status of 'Wizard' in the game, and now held the power of 'life and death' over lesser players. He recalled his more lowly state:

> 'I remember the first time I was killed on MUD—it was deliberate. I was in tears. I really knew what it was like to be dead, the simulation was so real.'

Although the majority of the players tended to be young men, a female interviewee was also a regular MUD player, and experienced the same stimulation.

> 'You get very excited with adventure games. Your adrenalin goes up and you get very tense. It's fascinating that you forget that you are hunched over a computer and that others are—you feel you are all together in the magic land of MUD. It's a further extension from reading a book. It's

totally engrossing—the mind is focused on one thing and you don't notice anything else.'

She, like many others, was an avid science fiction reader. Adventure games were a logical extension of this interest, but in this case one could control the progress of the story and be personally involved with the characters; a far more exciting prospect than merely accepting the settings described by an author, however brilliant.

When not playing adventure games, this interviewee spent her time on the networks as a games advisor. When players came to an impasse in a game, she would offer the solution for that stage, allowing them to pass through that sequence to the next challenging puzzle to emerge, and she had become quite a guru to others less experienced in the art of lateral thinking needed to play such games. She had rarely done more programming than copying code from magazines, and stated that she only used the computer to play games and as a means of getting to know other people; *i.e.*, as an aid to communication. The impersonal nature of learning about others simply through their typed words, as well as having many advantages, could lead to misconceptions, however. She, a woman in her fifties, was being prompted to meet some of the young, male networkers at a computer show. Although never hiding her age she felt that they would be less interested in communicating with her after realizing that she was 'old enough to be their mother or even grandmother'. Through her knowledge she had earned their respect, but she felt that this could diminish with face-to-face contact; the mystery would vanish.

Others were able to maintain the mystery, however, as fantasy played a great part in much of their networking. The anonymous use of a modem enabled players to use 'personas', a self-styled name such as *Garth* or *Zork* often based on science fiction characters, in order to project the desired image of themselves. One could give the impression of youth, humour, strength, beauty or whatever in this manner. Some players used more than one persona, and one man admitted to using a female name at times so that he would 'be treated more gently'. Such sex-role experimentation was not very common on the networks, but there was a much repeated apochryphal story of one of the few women who played MUD who, after two of the players had fallen in love with her and tried to make direct contact, was found to be male. Being a woman, real or otherwise, did have effects upon the game, however. I was told that the MUD commands given would alter from the usual 'immobilize' and 'stun' to those such as 'fondle' and 'kiss', bringing new forms of fantasy to the game.

The ability to use a medium which could give one total anonymity had many advantages. Some stated that they were able to communicate with the types of people whom they would never meet in real life, and that it gave them the ability to experiment with various types of interaction which they would be too reluctant to use when physically in the presence of those people. Networking is more anonymous even than the communication between pen-friends, where one can be evaluated by one's handwriting and where photographs are customarily exchanged. If all communication is achieved via the keyboard one is accepted by one's typed expressions alone. Even one's spelling can say little as networkers often develop very idiosyncratic modes, such as 'da noo pooter service', 'he sez', 'abart time', 'oi'm reddy'.

Relationships, real and imaginary, have been started on the networks and some act as virtual dating agencies for people who may never meet each other. I was given a printout, down-loaded from a Californian network, which indicated just how far the network contacts had gone in the United States. This network was for 'hot chats' only, and the printout indicated that this had little to do with the weather in that part of the world. Although I did not gain the impression that things had gone quite so far in Britain, *Heart Search* and *Sympatico* both provide similar, if less torrid, services in this country. (One interviewee, an Explorer, was infuriated when his wife learnt how to access the networks while he was at work. He could not cope with the fact that she was having 'relationships with other men', while his own computing consisted of innocent programming).

Although the Networkers were shown to be less shy and more extrovert than other Dependents, it should not be assumed that they were having intimate 'conversations' with others via their modems, unless of course they were interacting with someone of the opposite sex. However frequent the contact, some did not use the experience in order to exchange any personal information about themselves, nor wished to hear such details from others. The networks were for the exchange of ideas and computer information only. One man described his personal communications over the networks as follows:

> 'I do use the computer mainly to communicate with others, but they could be another computer to me. (My wife) is interested in their ages, whether they are married, their jobs, etc. – I couldn't care less whether they were people or computers, but I am interested in meeting S. . . . and M. . . . because they are women.'

The fantasy aspects were obviously important for some of the interviewees and the networking systems served their needs in a harmless but effective fashion. Exhibiting similar qualities of fantasy and excitement, 'hacking' into other computer systems was also a popular activity for some of the Networkers, and was often undertaken in cartels rather than in isolation. (See Cornwall, 1986, for a description of the equipment and techniques necessary to become a successful hacker).

All mainframe computers with multiple users employ some form of internal security system, encryption codes and passwords to prevent unauthorized people from gaining access to their databases. Even the networks, such as *Prestel, Telecom Gold* and *Micronet 800*, to which many subscribed, have private 'mailboxes' to be accessed by one person only. The philosophy of the hacker is that any code or password devised by man can also be decoded by man. The amateur hacker is not primarily interested in undertaking any mischievous or criminal damage when he has broken into a system, but merely wants to beat the experts and unscramble their code. Most of the Networkers had made at least some attempt at hacking, and some had become very adept and 'successful'. Many hackers know each other and pass on known access codes from one to another.

Numerical codes, which may be changed daily in order to inhibit the entry of the uninitiated, do not deter the hacker. With a microcomputer and a modem it is quite possible to generate random numbers which the computer can churn through until access to another system is gained. Once into the main system it is necessary to gain access to the files, and for this a password is needed. I was frequently shown, using just a small microcomputer and a modem, just how easy it was to access 'protected' systems. Many people seem very unimaginative when choosing their own passwords and it takes little skill on the part of the hacker to find a password in order to gain entry. Most of the following have been used successfully many times, 'hello', 'password', 'fred', '0000000' or 'ABCDEF', and one interviewee even managed to gain access to a large database by using the password 'hacker'.

Once into the files, the hackers often do no more than look around, read the private mail, or at most leave a witty message to let the other users know that a hacker has been in. One of my interviewees managed to hack into a live BBC television computer programme, causing much embarrassment to the presenter, and via the network was invited to take part in the following programme to explain how he had managed to gain access. This he did, and was filmed in silhouette to avoid identification. Most evade detection, but at least one other research participant had been identified and brought to court, while still another made a 'moonlight flit' with all of his computer equipment, believing himself to be discovered. Hacking is somewhat dishonest, but seemingly respectable businessmen, who in other circumstances would be appalled by overt theft or burglary, quite happily showed me their hacking skills. The morality seemed similar to that which believes it is acceptable to defraud the Inland Revenue but not to steal from a friend.

The Networkers' personalities, social lives and their computing activities did differ from those of the other groups. Their computing was seen as fun and games, and provided them with hours of entertainment. Although socially inhibited, this was not as pronounced as for the other groups and this, together with their less hostile and fearful attitude towards people, appeared to make them less likely to choose the solitary activity of programming as their main activity. Although they were no more likely than the other Dependents to share their computing interest with other members of their families, and still reported similar problems occurring because of it, their preferred activities did involve communication with others, even if only indirectly. They preferred to use the computer as a means of communication, although the personal interactions made usually remained secondary to the activities of games-playing and hacking. The computer was not an end in itself but merely a means to an end; to gain knowledge, for amusement, and as a challenge, even if dishonest, and as fuel for their intelligence and imagination.

There were no suggestions from the results obtained that the Networkers' personalities had in any way been affected detrimentally by their use of the networks. Conversely, there were suggestions that some had learnt to become more bold socially and that friendships had developed with people with whom they had communicated in this way. Just as C.B. radio had provided a communication channel for those who were lonely and shy, so too could computer networks. The advantage of the networks lay with the fact that mere use of the computer ensured that one was interacting with others who were also interested in new technology. This was of great importance to the Networkers who, like most of the other interviewees, often found it very difficult to share their hobby with others.

Brod (1984) believed networks to be the ideal medium for gaining understanding and companionship from like-minded individuals and discussed their use as dating agencies, and Frude (1983) believed that the isolated and the disabled were likely to benefit greatly from their use. The positive aspects of networking were also recognized by Boden (1984) and Kiesler *et al.* (1984) who believed that the use of networks could be liberating for the more socially shy, and because of the anonymity afforded could promote intimacy and reduce self-consciousness. In addition, Edinger and Patterson (1983) felt that the lack of social cues involved in the communication would make it more egalitarian, and allow the inhibited to become more assertive in their interactions with others. Perhaps the mere experience of using the networks to experiment with different types of interaction would carry over when in face-to-face communication with others, as one of the interviewees (a Worker) had in fact met his wife after getting to know her over the network, and had until that time little experience with women.

The games-playing and hacking undertaken by the Networkers also seemed to encourage social interaction, although games-playing has been reported to have detrimental effects upon people (see Weizenbaum, 1976 and Favaro, 1982). However, when Gibb *et al.* (1983) undertook research to determine whether this were so, they found no indication that games-playing encouraged social isolation even when playing solitary games, as ideas and games were often exchanged with others. They found that their 280 subjects showed less obsessive-compulsive behaviour than given for the established norms. They concluded that this was due to the fact that the games required flexibility of responses, not mastery of a rigid response pattern; a fact which is especially true of adventure games such as MUD which were of especial interest to the Networkers. Nicholson (1984) recognized that the instant feedback and feelings of control engendered when playing games could be psychologically very rewarding, and give an increase in self-esteem and belief in one's own competence and abilities. Similar theories were expressed by McClure and Mears (1984), who also showed that computer games were especially enjoyed by young males who liked challenges, (and in addition liked science fiction), and that adventure games were of greater interest than arcade games to those with high intelligence.

Although the results showed that the Networkers rarely undertook any programming (the average was 22 per cent of their home computing time), when they did it was invariably of an exploratory and self-educational nature, as undertaken by the Explorers.

All of the activities enjoyed by the Networkers confirmed their attitude that the computer was seen as a toy and computing as a fascinating hobby. Their activities were undertaken for fun and intellectual stimulation, and had the added advantage that it enabled them to increase the range and scope of their social contacts.

The Workers

The Workers formed the smallest of the three groups. These people, all of whom were male, concentrated upon writing programs with a specific end-product in mind. Although some had enjoyed programming in an exploratory manner during their early experience as computer owners, this self-educational type of computing was no longer paramount; computing had become a means to an end, and not merely an entertainment or an intellectual challenge. The results in Table 9.4, together with the interview data, combined to give a composite picture of a group of people who obviously enjoyed computing, but who tended to take it (and life) seriously rather than frivolously. To them computing was not merely a hobby, but a means of realizing ambitions by using the computer as a tool rather than a plaything.

In many cases their ambitions appear to have been fulfilled as several of this group were self-employed, with jobs for which computing was central; this was an aspiration of many of the other Dependents. Even a schoolboy within this group had already sold programs which he had been commissioned to write. He planned to gain a degree and go to America as fast as he could, 'to go into data communications, take over Houston and amass lots of money.' The Workers showed themselves to be ambitious careerists who were less likely to differentiate between work and play than the other interviewees, and while at work felt they

Table 9.4 Results differentiating the Workers from other Dependents

Results which differentiated the Workers from others		p
Questionnaire and interview data		
More likely to have attended public/boarding school		*
Less likely to have enjoyed sport at school		*
More likely to read non-fiction as a child		*
Less likely to differentiate between work and play		*
More likely to have autonomy at work		*
More likely to be self-employed		**
Mothers more likely to be described as housewives		*
Owned more computers		**
Value of hardware approximately 3 times more		**
Software more likely to be bought that copied		**
Spent more hours computing at work		*
Spent more hours computing in total (mean 55·1 hrs/week)		*
Partner's assessment of time spent computing more		*
Less likely to find it difficult to stop when computing		**
Spent less percentage programming time in exploration		*
Computing more likely to be for a useful end-product		*
Programming usually pre-planned not hands-on		***
Programs considered to be well-written		*
More likely to feel creative when computing		*
Computing seen to have broadened intellect		*
Less interested in computer games		**
More likely to disapprove of hacking		*
Leisure activities inventory		
Workers less interested in	Adventure games	**
	Computer games	*
	Computer hacking	*
	Reading fiction	*
	Reading poetry	*
	Writing stories/articles	**
16PF Questionnaire		
Factor G Expedient : Conscientious	Workers more conscientious	*
Factor Q3 Self-conflict : Controlled	more controlled	*
Factor Q4 Relaxed : Tense	more relaxed	*
Self Report Scale		
Workers saw themselves as	less emotional	*
Ideal Self Scale		
Workers wished to be neither shy nor out-going (rest out-going)		**
Attitudes towards computers		
Workers saw computers as	less interesting	**
	less risky	*
Attitudes towards people		
Workers saw people as	less critical	*
Attitudes towards mother		
Workers saw mothers as	less conformist	**
	more frank	*
Attitudes towards father		
Workers saw fathers as	more distant	*
	less loving	*
	less out-going	**

experienced and had achieved more autonomy. Although the Networkers and the Explorers who used computers at work were usually unable to indulge in their preferred computing activities until they were at home, this conflict did not arise for the Workers. Their work and play activities were compatible and had taken over much of their lives.

Although the results showed that the majority of Workers' computing was for serious ends, it should be remembered that they still considered themselves to be dependent upon computers, hence their participation in this research, and that they and their partners considered them to be no less dependent than the Networkers or the Explorers. In some ways one could suggest that they were even more dependent upon computers than the other two groups. Not only did they devote more of their work time to computing, as would be expected, but the estimates of their total hours per week and their partners' assessments of the time spent computing all exceeded those from the other two groups. They committed more time and also made greater financial investments in their computing. They owned more computers, and the average estimates made of the value of their hardware was over £4000, almost three times that of the other Dependents.

The methods of programming used by the Workers were also found to be significantly different from the other two groups. Although similar proportions of all three groups had taught themselves to program, the Workers' need for a usable end-product appeared to have made them more aware of the need for some element of structure and order in their work. Their programming tended to be pre-planned, even if only roughly, rather than hands-on, and their subsequent programs were considered in the main to be well-written. The Workers felt more creative than the other groups when computing, and also described computers as less risky and less interesting than the Networkers or the Explorers; attitudes which reflected interests more concerned with the end-product of the work than in the process of programming. Because of their pre-planning this group also did not find it as difficult to stop computing as the other groups, although they were equally susceptible to the experience of losing track of time.

This group frequently used commercial business software, most of which they purchased or legally acquired with their hardware. Unlike the other groups, the Workers tended to see the pirating of software as immoral, perhaps because they were software writers themselves. Similarly they were less likely to approve of hacking, although equal proportions of all three groups had attempted the activity. Lacking the attitude that computing was an entertainment and an intellectual challenge, neither these 'schoolboy pranks' nor computer games were of great interest to them.

The Workers gave the impression of a group of people for whom the work ethic was dominant, an attitude which appeared to have developed early in life. When young they were serious, academically-minded children who neither appreciated nor needed the physical escape of sports, nor the mental escape of the fantasy of science fiction so popular with the others, preferring to concentrate their reading upon non-fiction.

All the members of the group stated that their schools had strong academic biases. From such schooling, which none seemed to find very enjoyable, most had been very high achievers, with those over school age all entering some form of tertiary education. As well as attaining a good crop of degrees, two of the group had also undertaken extra business courses to further their ambitions. At school the Workers tended either to be outsiders or 'in charge of things', and like the Networkers often took posts of responsibility at school, although probably not for the same reasons.

Their fathers were described as solitary beings and appeared rarely to interact with their sons. They were reported as unfeeling, distant and cold, with two of the Workers implying that they despised their fathers; one for the cruelty shown to his mother, the other for neglecting the family by never being at home. None wished to resemble their fathers in character, and disliked any attributes about themselves which they saw as similar. The results from the scales confirmed that their fathers were indeed viewed as less loving and out-going and more distant than those of the other Dependents. Adulthood had not altered these attitudes, although one man now felt more powerful than his ageing father whom he now pitied.

While all but one of their fathers were professionals, often in business, this was the only group for whom most of the mothers were described as 'housewives'. Although their mothers were viewed more positively than their fathers and were described as more caring, only one spontaneously described his mother as loving. This group seemed to be particularly unsure of their parents' love for them, as was illustrated by the following quote:

> 'She's caring, but not a motherly mother. I never took their love for granted. I don't like to have emotion in my life, so I'm not sure if she's loving.'

This comment seemed aptly to describe not only their mothers but themselves. This group had experienced social difficulties from an early age, and most saw themselves as solitary even when adult. This on the whole did not seem to affect them adversely as they felt they earned people's respect by their achievements. It was therefore noteworthy that they found people to be less critical than did the other two groups. Few seemed to recognize emotional aspects of their own personalities, which either seem to have been suppressed or never developed, and the scale results confirmed that they rated themselves as less emotional than the other Dependents.

Most members of this group seemed to find it especially difficult to offer more than the briefest answers to any personal questions. Not only did the Workers find it difficult to describe themselves, but they tended to concentrate upon the task-related aspects of their personalities when asked to give an overall impression of themselves.

> 'I don't know, don't think I know. I'm fairly independent and can handle most things that get thrown at me, and am quite happy doing it.'
>
> 'I've a good mind and I tend to succeed in whatever I set out to do. I'm strong and determined—a man who drives hard and is proud of it.'
>
> 'I'm a perfectionist. I'm non-introspective, I can't answer the question. I don't categorize myself in that sense.'

The results from the 16PF, which showed the Workers as significantly different from the Networkers and the Explorers, revealed them to be more conscientious and controlled, but also more relaxed, with higher scores on factors G and Q3, and a lower score for Q4. Cattell *et al.* (1970) described those with elevated scores on factor G as persevering, determined, responsible, emotionally disciplined, concerned with moral standards and rules, and dominated by a sense of duty; all apt descriptions of the Workers, with the latter not only reflecting their responsibilities to work but also within marriage.

> 'It (marriage) provides me with a necessary requirement to fulfil, by maintaining an establishment. I have another human I'm responsible for. I have a requirement in life, this gives me stability.'
>
> 'As a family man I very conscientiously try to be loving, I'm working on that all the time. Frictions that do arise stem from me playing the role of the ideal man, but woman's ideal man is different.'

Raised scores on Q3 show characteristics not totally dissimilar from those of factor G. Such people are shown to be self-controlled, objective, decisive, persistent, conscientious and effective leaders. Cattell *et al.* (1970) stated that people with such high scores also make a definite effort to conform to 'clear, consistent, admired pattern of socially approved behaviour'; as was illustrated in the last quote from a man striving to conform to his own ideal, regardless of the consequences.

The lower scores for Q4 showed the Workers to be less tense and frustrated than either of the other two groups, no doubt partly due to their increased levels of autonomy in their work. O'Halloran (1954) and Wright (1955) [both cited in Cattell *et al.*, 1970] found low scores on Q4 to be correlated with high rather than low achievers of the same intelligence level, but negatively correlated with trial-and-error performances. This would concur with the Workers' need for productive end-products from their computing and their disaffection with hands-on, exploratory programming.

Although more socially inhibited than the Networkers, the Workers could be seen as successful in the worldly sense. Their hobby had been turned to their advantage, giving them the prestige, status, worth and independence they desired. They were idealistic realists for whom fantasy seemed to play little part in their lives, and they had been able to realize their ambitions through hard work and effort. Their manner of living and their work were taken seriously, and to them the computer was a means to an end, not an end in itself or a toy.

Brod (1984) described this type of computer 'workaholic' as obsessed with the need to shape the world by their own will-power and effort. Although he realized that their attributes were highly prized in the western, industrialized society, he believed that such activity was characterized by excessive rigidity, over-specialization and emotional detachment; factors which were in keeping with the Workers' personality profiles. He also suggested that they tended to be people who are full of self-doubt, latent hostility and feelings of personal guilt. These aspects may have been present, but were not readily accessible to observation from the measures used in this study.

There was little doubt that Workers' detachment and need to control their emotions did cause problems to the partners of some of those interviewed (as will be discussed in the following chapter); however the effect of the Workers' computing upon others did not appear dissimilar from that caused by the other interviewees. In addition, the use of the computer was not thought to have been the cause of these difficulties or personality traits; they had always been apparent, but computing did appear to give much greater emphasis to them.

The Explorers

The Explorers formed the largest group to be isolated from amongst the interviewed Dependents according to their preferred computing activities. They differed from the other interviewees in fundamental ways, and the results showing significant differences are summarized in Table 9.5.

Although both the Explorers and the Workers devoted 66 per cent of their computing time to programming, the manner in which these two groups programmed differed significantly. While the Workers concentrated upon producing useful end-products, the Explorers devoted the majority (72 per cent) of their programming time to tasks of an exploratory and self-educational nature. This type of programming had been undertaken by many of the Dependents during their initial experiences with computers, but because of differences in needs and personalities the Workers and Networkers had in the main forsaken it in favour of other interests. However, even after many years of computer ownership the excitement of exploratory programming had not waned for the Explorers, and this group could be seen to epitomize the characteristics of those described as computer junkies and 'compulsive programmers' (Weizenbaum, 1976). [The results revealed that both the

Table 9.5 Results differentiating the Explorers from other Dependents

Results which differentiated the Explorers from others			p
Questionnaire and interview data			
Less likely to have enjoyed primary school			**
Less likely to have taken posts of responsibility			*
More likely to live alone			*
More likely to spend free evening computing			*
Less likely to feel that job makes use of their talents			**
Preferred activity: exploratory programming			***
Greater percentage programming time spent in exploration			***
More interested in understanding the hardware			*
Networking less likely to be spent in personal communication			*
Programs rarely considered useful (in comparison with Workers)			*
Program hands-on, not planned (in comparison with Workers)			***
Programs not well-written (In comparison with Workers)			*
Leisure activities inventory			
Explorers more interested in athletics			*
16PF Questionnaire			
Factor C	Affected by feelings:	Emotionally stable – Explorers less stable	*
Factor O	Self-assured : Apprehensive	more apprehensive	**
Factor Q4	Relaxed : Tense	more tense	*
Factor QII	Low anxiety : High anxiety	more anxious	**
Factor QIV	Subduedness : Independence	more subdued	**
Self Report Scale			
Explorers saw themselves as		less ambitious	***
		less fulfilled	*
		less stable	*
		less relaxed	*
		less trusting	*
Ideal Self Scale			
More Explorers wanted to be		extroverted	**
Attitudes towards computers			
Explorers saw computers as		more straightforward	*
Attitudes towards people			
Explorers saw people as		more distressing	**
		more judgmental	***
Attitudes towards mother			
Explorers saw mother as		more anxious	*
		more shy	**
Attitudes towards father			
Explorers saw father as		more cheerful	**

Explorers and the Networkers were exploratory programmers, although the Networkers showed little interest in this activity only spending 22 per cent of their time programming. Consequently, the descriptions of differences in programming methods concentrate upon comparisons made between the Explorers and the Workers.]

The Explorers interviewed revealed that their yearning was for comprehension, understanding and mastery of the inanimate world and this was reflected in their computing, which was a source of intellectual stimulation and challenge. Although bubbling over with ideas, many of their computing aspirations appeared to be impossible to achieve. Their programs were often too advanced for the capacity of their equipment or would take years to complete, but this in no way detracted from the liberating excitement of dreaming that one day they would complete the 'ultimate program'. A program to allow one to control the computer using brainwaves alone was the aim of one of this group. The Explorers' programs were usually without useful end-products and many remained unfinished. Most were developed in a hands-on fashion without pre-planning, and were not considered to be particularly well-written. Programs which were able to be completed were undertaken to demonstrate the power of the computer, determining π to 10 000 decimal places for example, or in order to solve a particular algorithm (which itself might be intrinsically interesting, but often not extrinsically useful in a practical sense).

Debugging programs, frequently a very time-consuming activity when programs are poorly planned, was often described by the Explorers to be the most interesting part of programming and infinitely more exciting than writing the initial code. (This interest had initially been voiced by the undergraduates interviewed and was shown to differentiate significantly between the Dependents and the Owners.) The Explorers gained the greatest enjoyment from debugging, emphasizing their desire to investigate every aspect of computing with little regard for efficiency or the completion of tasks. Many programs were written in a few hours, deliberately leaving the Explorers with weeks of work to debug and refine them.

> 'Debugging is the enjoyment of programming. I like changing and adapting things—that's the only way to do it. I *ad lib* everything, I *ad lib* life though. If you haven't planned it and it goes wrong, so what? I get a high when it works.'

> 'The program can be a work of art which no-one else can understand, but it does something beautiful. These are the ones, they solve a problem in a very elegant, unique way, or are quick and powerful, or it does something you could never imagine it could. Bugs can produce more beauty—human bugs can produce problems.'

> 'I used to use pen and paper at first, then I composed straight in. It was a way of getting as much reaction as possible. I tried to push the program to its limits, but I wanted to find faults. I was always having to go back—I wondered if this was often deliberate. I never left it alone, it was never perfect, perfection meant it was finished.'

> 'My programs are still all half written. I like to learn, I keep doing things to them to find out more. I've written an adventure game, completely different from any other. It's never been finished. It's non-mappable and completely random and has good humour in it. I've been working on it for five years.'

> 'I spend a lot of money to have the right bits and pieces, so it looks good aesthetically. It's the physical feel and the interactiveness. I'm a curious person, I've got to find out what's going on—knowledge is power—I'm a guru when I'm computing. I do look forward to coming home to it. It's the power of being able to do it myself, it's mine, I can make it do what I want it to do. It's only limited by my capacities, and mine will run out first.'

All aspects of their computing were exploratory, not just their programming. The Explorers expressed a greater need to understand their hardware than either of the other two groups, and some used their machines with 'their guts exposed' so that adaptations could be affected easily. The majority of this group considered hacking to be 'fair game' and an acceptable way of using their investigative skills, but unlike the Networkers this tended to be a solitary rather than a shared activity. Networking was undertaken, but the Explorers were found to be significantly less likely to spend their time in personal communication with others, although some had tried to share their ideas with others: 'The first time-frame meeting was weird. All these weird, introverted people not talking to each other.' Most of their software was pirated and the Explorers spent a great deal of time breaking the protection codes of games and other commercially written programs, as well as investigating the software integral to the machine.

> 'The computer almost appears to be alive. I don't really think it is a silicon life form, but it's a very intricate machine which must have a logical base. I just like finding out about what makes it work. I can find out about the person who wrote the operating system, I think he was as obsessive as I am.'

> 'I enjoy poking around in the memory, it does some alarming things. I spend a lot of time mindlessly poking and peeking. I make little loops in the graphics and sound programs so that it will repeat itself. In ten lines you can have a fully working program. The feeling of power is gratifying. You always know there is something else you can do—it's continual growth and mastery. It's the ultimate creative device, and it's unlimited in its applications. It's something on which I can be creative and practise other people's ideas.'

> 'I've spent hours cracking pirated software systems. I like cracking into games especially and getting them to work better, and removing the bits of programs which are supposed to prevent piracy.'

> 'They cry out to be used optimally. Some people use it as a tool, it's convenient and does a job, but then there's the programmer who is locked

into a cycle. It's like a vortex, you get locked in. It's quiet on the outside and weak—here the magazine readers and the club members drift in and out at will. But in the middle it's turmoil—you get locked into the stream, you get sucked in. The further you get sucked in the more difficult it is to get out, but you don't want to—it's so exciting.'

While the Workers and the Networkers viewed the computer as a tool, as a means to an end, the Explorers saw their programming as an end in itself, and described their computing as an intellectual challenge, and as fun and entertainment (see Table 9.2) They tended to revel in the jargon and were not inhibited by the poor manuals usually provided with microcomputers. They wished to learn using their own methods in order that they could fully understand and exploit the systems to the full. Although interested in the technology behind menu-, mouse- or voice-driven computers, all methods by which computers may become more straightforward to use, they had no desire to transfer to this type of equipment. They preferred the computer to remain intriguing, perhaps to prevent the common man from acquiring their skills, although this was never expressly stated. The greatest insult was to treat the computer as a mere problem-solving tool for the masses. Such protective views were similar to those of the C.B. radio operators of a few years ago, who lost all interest in them when they became legal to use and were bought by the uninitiated.

Although absorbed by their hobbies, the Explorers seemed to have devoted less energy towards their academic careers and professions than the other groups. Some felt that they had underachieved academically, even though there were no significant differences between the three groups for their highest educational attainments. Perhaps through lack of ambition, they were also less likely than the other groups to be in jobs which they felt made good use of their talents. The impression given was that some of this group tended towards laziness and apathy and rarely seemed to strive to achieve, unlike the other groups.

'Academically I was an underachiever. I was lazy and other things took my interest. I could have done far more with my life.'

'There were so many other distractions at university. I didn't work hard enough and I was gradually sinking lower in reference to my peers and getting fed up. I spent my time in the Science Fiction and Science clubs.'

For a few of the Explorers this general sense of apathy had pervaded all areas of life, including their past hobbies, and in spite of their present dedication to computers some seemed unable to justify or understand why computing had become such an intense interest.

'I've never found anything to really be involved with until now. I'm totally involved, I even resent having a bath now.'

'I have no idea, I'm very surprised, I can't explain it. When you start messing

> about you become interested. I got hooked into breaking into programs, I can now copy the uncopyable.'
>
> 'I can't justify what I do to be quite honest. I've wondered about that for some time. I think people with inadequate personalities use it as a compensatory mechanism.'

Not all showed such general apathy however, and their computing was shown to be the most important from a series of absorbing hobbies. For some their leisure activities, where their efforts were self-directed and controlled, appeared to be the main area in life for which they had consistently been inclined to devote their energies.

> 'Previously I spent all my time in the garage, designing and building a car. I used to make telescopes, and joined an astronomy club. I always spent a lot of time on my hobbies.'
>
> 'When I get a hobby I do it properly. I spend a lot of money and time. I've been into astronomy, science fiction books, Dungeons and Dragons, lead figures, HiFi. It sometimes worries me that I don't do enough work.'
>
> 'I was very seriously into music, playing music, model railways, making explosions—cannons, fireworks, into picking locks at University, electronics, practical building things. At four I wanted a blow torch and a computer—I didn't really know what it was—I was into crystal sets and geometry for fun.'
>
> 'Philosophy was my major interest for 10 years, and C.B. until it was legal. I've always been heavily into my hobbies, I'm definitely not a jack-of-all-trades type. I have to do things properly.'
>
> 'I've always got quite obsessed with my hobbies, mainly photography and developing and electronics, and I couldn't live without music.'

Computing was considered to be the best of their hobbies and 'in a different league'. The fact that the Explorers were significantly more likely than the other groups to describe the computer as a friend or companion may have accounted for this. The computer was felt in some way to have a personality of its own, and it was extremely interesting to note that those who personalized and anthropomorphized their computers always referred to them as male (from whichever group they came), no doubt because they felt the qualities of the computer reflected those considered to be masculine traits. This differs from other artefacts such as boats, aeroplanes and cars, which if ascribed a sex are traditionally seen as female. This masculinization of the computer occurred for both men and women who personalized their computers, with one woman describing it as 'the man in my life'. There was one interesting elaboration on this theme. 'It's my mate Micky. I always think of it as more masculine in its characteristics, but I always refer to my dual disc-drive as female—she's lovely'. This was said in all seriousness. His intelligent, logical computer was male, while his passive, receptive disc-drive was female. Whether or not he realized the implication of this statement was not pursued.

In spite of the masculinization of computers, two men deemed them to be sexy and sensual. 'They look nice, they are sexy, that's true, they are styled to look nice. When a new machine arrives you stroke it and touch it.' In many instances it was also apparent that the computer was considered as an extension of themselves, and there appeared to be a narcissistic quality to the interactions. 'It's me, it's just me in the machine—it's got my personality. I own it, all the stuff in it is mine'.

The easiest relationships are perhaps always with those who reflect our own opinions and attitudes, therefore if one is the initiator and the designer of the human-computer interaction, the 'relationship' would be seen, even if subconsciously, as a reflection of oneself and therefore comprehensible and accepting. For this reason, perhaps, the Explorers were more likely than the other groups not only to see the computer as a friend but also to describe it as easier to cope with than people, as an escape from social relationships or as a substitute for other less satisfactory areas of their lives. The need to have control over some aspect of their lives seemed especially important to those who found human relationships particularly difficult.

> 'It's just the sheer feeling of control over something that doesn't fight back.'
>
> 'The ultimate goal has got to be a machine that is more intelligent than humans. It is a friend. I'm in competition with it, I have to master it and make it do anything it's possible to do.'
>
> 'I use it as an escape, that's the major reason. I have a need to be wanted, loved, etc. It does provide compensation. You're in charge, in control, there are no real surprises, you can anticipate, you're never in control with human beings.'
>
> 'The satisfaction is in doing something well and having the machine at your control, and doing exactly what you want it to do without any argument. It's a friend, perhaps it is. It talks back to me, asks me questions, tells me that it doesn't understand me. That could be part of it.'
>
> 'I don't like being beaten. I make it do as it's told—having an idea and making it come true. I can't do that anywhere else in life.'
>
> 'I'm hooked probably because a machine is a lot easier to deal with than people. If you tell it what to do it will do it. People are unpredictable in how they will react . . . I have control over what it does. Given you know the right things to do you can get it to do anything. There's a set of rules to follow—people are not like that. "Play safe, make friends with a computer" I say. People are not like machines, they have a mind of their own and feelings—the computer will not leave home if you abuse it.'

Many of the Explorers found it very difficult to socialize with others, often describing people as incomprehensible and frivolous, confining any social activities to their immediate families or to a few computer friends. Intimacy was described as difficult, with some lacking the comprehension of what was demanded of them in social situations. Some as a consequence led very solitary lives, which they found preferable to the strain of making the effort to conform with what they considered to be socially acceptable behaviours. This strategy applied early in life as significant results showed this group to have been less likely to have enjoyed their primary school years or to have taken posts of responsibility while at school, suggesting long-standing difficulties with relationships.

The Explorers and the Workers were shown to experience significantly more social problems and to be more introverted than the Networkers. However, while the Workers found stability, prestige and control through their work and status, this did not apply to the Explorers who exhibited different personality traits. The results from the Self Scale showed the Explorers rating themselves as less ambitious and fulfilled, as well as less stable, relaxed and trusting than the other two groups of interviewees. Such results indicated that a certain degree of tension and frustration was present in their lives, perhaps caused by their poorer interactions with people whom they also saw as more distressing and judgmental than the other groups.

The results from the 16PF seemed to confirm these hypotheses (see Figure 9.1) As well as being as shy and introverted as the Workers, the Explorers were also found to be less emotionally stable, and more apprehensive, tense, anxious and subdued than either of the other two groups. It was pertinent that unlike the Workers, who showed the lowest scores for factor Q4, (relaxed : tense) the Explorers showed the highest. This factor has been shown to distinguish between the high and low achievers of the same intelligence levels and is positively correlated with good performances in trial-and-error tasks (O'Halloran, 1954). In this instance such interpretations of the 16PF results almost perfectly mirror the self-reported differences in achievement and computing activities between the Explorers and the Workers.

Cattell *et al.* (1970) found it was often difficult to separate low scores for factor C from high scores for factors O and Q4, all shown in the mean profile of the Explorers' results, and all of which were instrumental in producing the high scores for the Second Order factor QII, anxiety. From the interpretations of Cattell and his colleagues, people with the above traits tend towards moodiness, guilt-proneness, inadequacy, instability, irritability, and high levels of frustration. Such a person is described as 'easily annoyed by things and people, is dissatisfied with the world situation, his family, the restrictions of life, and his own health, and he feels unable to cope with life'.

Some of the Explorers had experienced trauma in their lives and most did appear to be frustrated with certain aspects of life or themselves. Their relationships with their parents had frequently been very poor, once again with fathers being shown to be cold and distant, often dominating both their wives and children, and with mothers shown to be significantly more anxious and shy than those of the other groups. Such backgrounds seemed to have had a profound effect upon their later dealings with people and upon their views of themselves, and some of the interview data is given in full to illustrate these aspects.

> 'I'm quiet, somewhat introverted, a bit rigid in views and attitudes. I'm searching for something, but part is resisting the search. I have potential and abilities to do amazing things, but there are severe blocks especially with relationships, trying to be friendly, etc. I'd like to be more out-going, not manipulated by society. I'm not contented, I have unreasonable expectations in relationships, I want things that others are probably not capable of giving . . . I used the computer as an escape, I was aware that I used it as an escape. They are not addictive *per se* . . . It replaces and compensates for defects in my own make-up.'

> 'I'm a failure. I have a failed marriage, a poor degree and I didn't fit my first job, although I'm quite clever at what I do. I'm somewhat inept socially, I can't believe that others would want to see me. I'm more tolerant than others, but when I hate I hate passionately . . . It was not a happy childhood, very working class. He was a drinker, she was a nag, insensitive to others, selfish, not very tolerant. I'm an oddity to my parents.'

> 'My father was a bad-tempered, weak character, he took it all out on his family. My mother was shy and retiring under my father's bad temper. I want to be anything but like him; I'd try and stop myself if I saw myself becoming like him.'

> 'Both my parents were unemotional, not physical. I didn't get on with my mother. She was someone I could shout at, she was always telling me what to do, but I realized I could use her to get what I wanted. . . . . I'm very confused by what it all means—I'm afraid, curious about dying. I think I was born middle-aged. I'm unambitious in the conventional sense, but I want to do something. I'm intelligent, but don't know how to use it. More ambition would give me more direction. I think people tend to avoid me, they think I'm a weirdo—perhaps I am, I'm unconventional.'

> 'My father was very Victorian, he expected instant obedience and had hundreds of scatty rules. We had to pick 20 weeds each Sunday! I always thought I was adopted—the boys were more important to the family. He was terribly strict, away a lot from home, an egoist. I couldn't stand him. He made me repressed and timid. He was always bossing me about, very dominant, very jealous. I never got close enough to him to find out who he was . . . (My mother) was always cheerful, we were good friends and very close. She would do anything for me, but she kowtowed to my father. She was a very anxious woman. I was very resentful—was made deceitful. I was so dominated. I resented working for him, but I was brought up to be a dutiful daughter . . . I was very unhappy when I first married, my husband was so selfish, I don't know . . . very moody, he was terrible with money, I hated having to defer to someone else . . . We divorced, he had another woman and I left the lot of them (including three children) . . . Being divorced has hardened me. I can't bear most men. I don't even kiss my own children, I don't display emotion. I think I'm a leader, I can't

> bear the fuddyduddying of everyone else. I can't sit back if something needs saying, I'm bossy. I'm as hard as nails, for years I couldn't cry. I can't get het up when someone old dies, we all die.'

> 'I went to 14 different schools, my father was in the army. He was brought up by his mother as his father died when he was young. He was very strict, hard, but good. There was a lot of physical pain, I used to get beaten a lot by one or other of them. Being as intelligent as I was, I think things should have been explained to me, not beaten in. My mother had an Irish temper . . . I was a loner. I'm a very strange person, very different. I didn't feel I belonged to the human race. I was superior generally, except physically. I expect the worst and I'm never disappointed by not getting what I wanted . . . but I'm never really contented. I'm self-centred, I find great difficulty in reading other people. I'm straightforward and honest, but they think I'm rude.'

Not all of the Explorers reported overtly traumatic backgrounds, but most had experienced some form of distress in their lives. A few described themselves as having neurotic tendencies or to have experienced bouts of depression at some stage in their lives. Several had had long periods away from school, through illness or accident, a few had experienced severe emotional distress as the result of broken relationships, and some lived in tension-filled marriages. Perhaps all such events would be inclined to make them less trusting of humans than either of the other groups, leading them to use the computer as an escape into intellectual activities. Computing had become an end in itself, providing them with an outlet for expression in an interaction which was stimulating, but safe and non-threatening.

The majority of the members of this group exhibited the characteristics of the schizoid personality type, and in most cases could be seen to have childhoods which could have induced this state. Their inability to form close, sharing relationships had, as a consequence, left some feeling inadequate in a world where poor personal interaction is considered to be unacceptable. The advent of the computer had managed to alleviate many of the problems they had previously experienced. Interaction with it was rewarding and satisfactory, and the stresses of their earlier lives had in most cases been removed as they no longer felt impelled to live a life which was alien to them.

Similarities may also be drawn between the Explorers and those with high achievement motivation. Vidler (1977) described such people as working for intrinsic rather than extrinsic rewards and for the opportunity to achieve excellence rather than prestige, both factors which differentiated them from the Workers. High achievers also desire personal responsibility for their efforts, and need to control their own destinies without recourse to the opinions of others. In addition, as time is felt to pass so rapidly for them, they feel there is not enough in which to do everything they wish. They desire challenging goals with uncertain outcomes and need immediate, regular and concrete feedback in order to monitor their efforts. Such descriptions almost perfectly mirror the Explorers' activities, and Winterbottom (1958) saw high achievement motivation, particularly prevalent in males, being caused by parents who set high standards and who reward children for the mastery of tasks.

There was little doubt that the computing activities enjoyed by the Explorers closely resembled those described in the literature by authors who had expressed opinions about computer dependency. Weinberg (1971) had experienced the excitement of programming and seemed to understand how people could become hooked on the activity, stating that 'programming itself . . . is the biggest motivation in programming'. Similarly, Shore (1985) described such an enthusiastic programmer as 'one who revels in marathon programming sessions that exploit every feature of a computer system, including features not intended or even known by the designers'; a depth of knowledge claimed by some of the interviewees.

Turkle (1984) recognized that the computer dependent person used the computer as compensation for dissatisfaction in other areas of life, but she believed this to be with politics and work, neither of which were mentioned by the interviewees. The outside world rarely seemed to impinge upon them, and the Explorers' jobs, although sometimes not believed to utilize all of their talents, were invariably fulfilling and did not cause them stress. Turkle also believed that computing was taken up during the teenage years by boys who had problems relating to girls, with the computer allowing them to achieve competency and control. Although there was some truth to this statement, the results from this research showed very conclusively that even when such hobbies were used as an escape or a substitute, they had been of great interest during the very early years of life and were not a pubescent occurrence.

Turkle divided the dedicated users into two groups, the hobbyists and the hackers, (using the term hacker in the global sense, rather than specifically to refer to breaking into other systems) and believed that they undertook different tasks for different reasons. She believed that the hobbyist was a risk-taker, who endowed the computer with magical qualities, and liked to make small changes to their programs to see what would happen. The hackers, on the other hand, she saw as needing reassurance by maintaining total control over the computer, by the use of assembler languages and mastery of the hardware. These two divisions were not at all apparent from the interviews. The Explorers showed both sets of characteristics very strongly; they were not mutually exclusive. They were only able to gain complete control over the computer by deliberately taking risks and observing the outcome, and one of the main satisfactions gained from their computing was that it allowed them to take risks within the confines of a safe environment, while simultaneously keeping control of the whole system. The computer and its behaviour, however bizarre, could not frighten or intimidate them; it was comprehensible and logical, unlike people.

Weizenbaum (1976) recognized the 'compulsive programmer's' need for power and control over the computer and that his programming was an end in itself, not a means to an end. However, recognizing the seductiveness of the positive and rapid feedback from the computer he believed it appealed only to the immature personality, who learns 'only its rudiments and nothing substantive at all'. Nevertheless, he did accept that the individual:

> 'is usually a superb technician, moreover, one who knows every detail of the computer he works on, its peripheral equipment, the computer's operating system, etc. . . . His main interest is not in small programs, but in very large, very ambitious systems of programs. Usually the systems he undertakes to build, and on which he works feverishly for perhaps a month or two or three, have very grandiose but extremely imprecisely stated goals.'

This was undoubtedly so. Many of the Explorers did find it difficult to explain exactly what they were aiming towards. The interviewee who wanted to be able to control his computer with his brain waves alone, via the use of electrodes, was not being overly ambitious nor ridiculous, however. Such systems are already in the experimental stage for use in fighter aircraft, with the aim of speeding the responses of the controls from the pilot's thought patterns rather than instrument manipulation. Neither this interviewee, nor any of the others, may produce a program which is workable and saleable, however their ability to think beyond the mundane is often highly praised in other areas of life. The extent of their ambitions was well expressed by one interviewee.

> 'What I really want is a computer built into my house. I want to build it and build my house around it. It will be a self-contained unit. When I find that hill and the money I'll start it. It will be able to run everything and any program from any machine. It's in my head, I know roughly what I need.'

The Explorers investigated each and every aspect of their own and other computer systems. They revised and rewrote commercial software having broken the protection codes, in itself a feat of computing skill, and many were experienced hackers. Their aim was to learn as much as possible about all aspects of new technology, whether by legal or illegal means. Levy (1984) described hackers as 'heroes of the computer revolution' and cited the Hacker Ethic which states that 'anything which might teach you something about the way the world works should be unlimited and total'. He and the Explorers believed in the total freedom of information and their right to have access to it. Levy recognized the need for some to learn about the world by dismantling and understanding its artefacts, and could only see this as a positive and enriching experience by giving one power and control over one's environment.

Although the critics seem to assume that hacking is a totally worthless exercise, there is little doubt that it gives great understanding about computer systems in general, and some past hackers have gone on to achieve fame in the world of new technology; Wozniak developed the *Apple* computer, Felsenstein the *Osbourne*, Smith the *MacIntosh*, and Greenblatt the LISP programming language. Even Weizenbaum found it necessary to concede that:

> 'Were it not for the often, in its own terms, highly creative labour of people who proudly claim the title 'hacker', few of today's sophisticated computer time-sharing systems, computer language translators, computer graphics systems, etc., would exist.'

Such success is rare, and I do not wish to suggest that all of the interviewees will achieve such heights, but one could assume that the drive and curiosity needed in order to explore and master the software and computer systems of others does expand the user's knowledge enormously, without which new developments would perhaps be slower to appear.

The Explorers felt themselves to be at the forefront of an exciting new age, and because of their computing expertize had more easily learnt to cope with their personality and social difficulties which had, until becoming involved with computing, often caused them stress and anxiety. Now their lives had meaning and purpose and this activity had provided the intellectual stimulation and security they needed.

Summary—All things to all men?

Although expecting differences in degree between the Dependent respondents, it had not been anticipated at the outset of this research that not all would be undertaking the same types of computing activities, and that quite marked differences would occur. The literature and the press reports all suggested that the 'computer junkies' were either programming or were hacking into other computer systems, neither with apparent extrinsic purpose. 'Addiction' to computer games had also been mentioned, and it was expected that more of the respondents would have indulged in games-playing activities to a greater extent. However, these descriptions proved insufficient to cover the activities observed amongst the interviewees. The playing of arcade-type games was rarely of interest to the respondents, although such games software was frequently hacked, explored and expanded upon. Nor was exploratory programming of interest to all of the interviewees, although most had gained their basic knowledge of computers in this manner.

The differences between what had been described in the literature and what was actually found when the area was fully researched was not thought to be due to inaccuracy on the part of other authors, but caused by the passage of time and the recent developments in new technology. When others were making their observations the home computer was rare and most computing was undertaken on mainframe terminals or in arcades, thereby limiting the types of activities able to be undertaken. The advent of the affordable personal computer had given the computer dependent person a new range of possible activities, and the development of the more sophisticated adventure games which were able to be played at home had superceded the 'shoot 'em up' type of arcade game. The power of the microcomputer and the availability of communications software had enabled the home computer users to develop their interests without restriction along new avenues.

Out of the confines of computer centres, where many of the interviewees had first developed their fascination with new technology, the Dependents of old were now able to have complete control over their equipment and activities. Those with the need to achieve and with the drive to turn their hobby into a lucrative, although still enjoyable, way of life were able to do so. Those whose main aim was self-education and exploratory programming were still able to pursue this activity, although they were also free to hack into other systems, contact other databases and individuals with a new freedom, or play with the new and exciting interactive adventure games. They were not restricted by the working hours or the time-sharing capacities of a computer centre, but were able to work freely at their own rates, pursuing their own interests which were shown to match their personalities.

The Networkers and the Explorers saw their computing activities as an intellectual challenge and as fun and games. Both of these groups enjoyed the naughty, illegal activities of hacking and pirating software and seemed free from the self-regulatory restrictions which govern the lives of most of us. The Workers, on the other hand, more conservative and emotionally controlled than the others, imbued their activities with more seriousness and rarely approved of these activities. The Workers needed to be seen to achieve and prove their prowess, although they did not share their activities with others. The Networkers shared their skills in friendly competitiveness with like-minded individuals, while the Explorers neither shared nor needed the approval of others. Although there were great overlaps in the responses from all three groups, each had its own preferred activities which were in keeping with their measured personality differences.

Even though the actual social lives of the three groups showed no differences, the Networkers, who were shown to be less socially inhibited than the other interviewees, used their computers to experiment with new relationships over the networks. They seemed more frivolously playful, and were more extrovert and confident than the other groups. Their family background experiences seemed happier and less traumatic than those of the other groups, although their parents' influences had still led them to be object-centred convergers. They were more likely to have had satisfactory social experiences with the opposite sex, perhaps influenced by their greater identification with their parent of the same sex as themselves giving them stronger sex-role models. Their computing activities were less expert than those of the other groups; they knew fewer languages and spent less time programming, mainly because they tended to use the computer as a communications device. However, they were still well and truly 'hooked' on their activities, and the MUD players' hours were perhaps more disruptive of their family lives than any of the other groups. Although appearing to have happier marriages, perhaps because of their greater social skills with the opposite sex, their computing activities still caused stress, were not shared with other family members, and still proved more fulfilling than their relationships.

The Workers devoted more hours than any others to their computing, but were perhaps more able than most to justify their activities, to themselves if not their wives. They appeared to be driven by a desire to prove their skills, which were more highly refined and professional than the other groups. This group were more likely to hold very negative attitudes towards their fathers, who it was revealed tended to be authoritarian and sometimes violent, less loving and more distant. It could be assumed that as a reaction to their fathers some appeared to have suppressed their feelings and had totally brought their emotions under control. They were not tortured by self-doubt; they had developed great independence and the need to succeed from an early age, and had single-mindedly pursued their goals. The work ethic dominated their lives and in many cases they could be seen to be conforming to the attitudes of this era by achieving the outward signs of success, even if this was obtained with little regard for others around them.

The Explorers' activities more closely mirrored those described in the literature. They tended to be solitary individuals whose computing was rarely shared, even with others of like-mind. Although stating that they were as contented with life as the other groups, the Workers were found to be less fulfilled and stable, and more anxious and tense. Their family backgrounds, which were frequently traumatic, could be seen to have been influential. The uncertainty of their early years had made them mistrust relationships, although unlike the Workers they tended not to blame their parents for the situations which had occurred nor did they feel they were in competition with them, rather they appeared to be victims of their early circumstances. Although a few gave the impression that their early lives were purposeless, intellectual activity had played a great part in the lives of most from an early age. They showed constant dependence upon the need to increase their knowledge throughout life, with the computer serving as the ultimate source of revelation. The constant stimulation and the pseudo-personal relationship able to be effected with the computer had provided them with an adequate substitute for human relationships. This had alleviated much of the stress which had dogged them throughout life, and by having little time for other activities, or the need to dwell upon their lot they had found contentment at the keyboard.

The results from this section of the research seem to suggest that the computer can be all things to all men. For those who are intelligent, curious and object-centred, playful or work-driven, who do not find social relationships stimulating and feel the need to turn away from the stresses caused by them, the computer provides the ideal medium for the satisfaction of their needs. However, because of its significant influence upon the lives of the Dependents, computing obviously had other effects, not only upon the Dependents but also upon those around them. The consequences, both positive and negative, arising from computer dependency are discussed in the following chapter.

Chapter 10
The effects of computer dependency: The advantages and disadvantages to the individual and others

Introduction

This research had been initiated following suppositions in the press and by various academics that dependency upon computers could have adverse effects upon the individuals concerned. The general impression given was that computer dependency was a 'bad thing' which ought to be 'cured'; no-one seemed to be concerned why it had occurred, although it was suggested that the introverted, the shy and those with social problems were most easily affected.

Attempting to determine why computing had come to dominate the lives of some people to the extent that it had, the majority of this study concentrated specifically upon the examination of factors which could be seen to be associated with the development of computer dependency. However, the effects upon the individual and others were also examined.

The advantages and disadvantages arising from computer dependency were determined from the correspondence received, from the interviews undertaken with Dependents, and from informal discussions carried out with some of the Dependents' partners. The results from this section of the research were therefore wholly reliant upon subjective assessments, as no empirical or comparative studies between them and the control groups were possible.

When viewing the results from this section it should be borne in mind that on no occasion was conflict found between the Dependents and their partners as to whether the Dependent person was 'hooked' upon computers or not. This applied whether the Dependent person had been self-reporting or whether the partner had made the first approach and subsequently persuaded their Dependent spouse to take part in the research. All involved appeared to recognize the syndrome, and when asked to assess the depth of dependency both the Dependent person and their partners offered very similar estimates (see Table 7.14). Nevertheless, differences frequently occurred when the Dependents and their partners were asked to assess the effects arising from computer dependency. While the Dependents stressed the advantages, their partners tended only to report negative effects arising from this activity.

Over one third of the interviewees had experienced some direct personal disadvantages, most of which concerned mild physical effects from the hours spent on the computer; detrimental effects upon their studies, work, finance or time; while a few reported negative effects upon their social relationships. For the vast majority of the interviewees these disadvantages were greatly outweighed by the many advantages which they had gained from their computing, but a few had to admit that damage to their marriages had been very severe and had affected them personally. The majority of those who had reported problems being caused to others concentrated upon the effects to their partners, although a few felt that friendships had also been damaged. Nevertheless, most of them remained personally unaffected by these problems, and the overwhelming conclusion drawn from the interviews was that computing was advantageous to the individual concerned but was often problematic to those who lived with them.

The results and discussions into the effects upon self and others are divided into sections to cover four main areas, as follows:

- Positive effects upon the Dependent individual;
- Positive effects upon others;
- Negative effects upon Dependent individual;
- Negative effects upon others.

Positive effects upon the dependent individuals

> 'He loved this feeling and continually sought to renew it, to increase it, to stimulate it, for in this feeling alone did he experience some kind of happiness, some kind of excitement, some heightened living in the midst of his satiated, tepid, insipid existence.'
>
> 'Siddhartha' (Hesse, 1973)

The positive benefits and advantages which were reported by the Dependents to flow from their experience with computers were both numerous and diverse, and many have been alluded in earlier chapters. There were intellectual effects, and most of the Dependents spontaneously stated that computing had enabled them to increase their analytical abilities and logical reasoning skills, and had increased their knowledge in various related technological areas.

> 'It sharpens your intellect and gives you training in lateral thinking. I like being challenged to think. You can get lazy in many jobs, but not with computing.'

> 'It increases your powers of reasoning, it exercises the mental faculties. I thrive on new things—it keeps my mind alive and sprightly.'

> 'It has made it easier to analyze things—it enhances your analytical mind in all areas. I think I have become more logical—it makes the brain more active.'

> 'I was never really interested in very much before. I'm, now interested in art and graphics—I try to get the perspective right, and I'm far more interested in writing now.'

Quite a number mentioned that their writing skills and abilities had increased dramatically, and some had written articles about computing which had been published in the computer press, something few would have attempted before the advent of word-processing facilities. The use of a spelling-checker on his microcomputer had enabled one man with dyslexia to gain far more confidence, to the point where he felt able to communicate on the networks.

As well as providing intellectual growth, experience with computers had dramatically improved the job prospects of many of the interviewees; from one unskilled man who taught computing in a primary school on a voluntary basis and hoped to develop this aspect further, to those for whom their knowledge of new technology had been directly attributed to their success in gaining well-paid employment or increased promotion within their companies. Many now felt a new sense of security, knowing that they had skills which were needed in the job market and felt assured that they would be constantly employed in this new growth area. Most loved their jobs and the challenges which came with them, and their depth of interest ensured that they would remain up-to-date and necessary participants in the world of new technology. Some of those still studying also found that their computing had helped their education in general.

> 'It's given me prestige and success and an advanced salary. It has improved everything.'

> 'It's kept me in jobs which are well-paid and interesting—computing got me the jobs, more so than my official education.'

> 'I have met so many people and I have a more interesting and satisfying job.'

> 'I got better grades than anyone else in computing, and both students and staff come to me for advice and I enjoy that.'

Another very significant benefit arising from the Dependents' knowledge and work experience with computers was the new prestige they had gained from their expertise, together with the pleasure obtained by their increased ability to offer advice to those less experienced with new technology. This proved very important to the Dependents, who for the most part had found little of common interest when in conversation with others. Now they were sought out by their colleagues and appeared to gain great pleasure by being able to offer their assistance.

> 'It's given me an understanding of a great many things—things which only a year ago would baffle me now I understand. I can take things on more confidently, and enjoy giving advice.'

> 'It is nice to know that you are competent at something. It's given me a lot of prestige—I love helping people if they are prepared to listen.'

Self-confidence and self-esteem had also increased dramatically, and most of the interviewees spontaneously used the word 'satisfaction' when describing the pleasure experienced when computing, many stating that it gave them a great feeling of achievement. In addition, computing offered many of the Dependents great fulfilment and a new, stimulating interest in life. Many had reported great stress when dealing with social situations in the past and stated that their computing had brought relaxation and a relief from this anxiety and depression, either by acting as an escape or a substitute, while others reported feelings of excitement and elation when computing.

> 'It's stopped me thinking about myself. It's given me something to do and stopped me being bored—I hate to be bored. It's the pleasure of achievement—that covers it all.'

> 'I used to suffer from depression—that was caused by life not computing—computing has helped it.'

> 'It's the power of being able to do it myself. I get a feeling of being relaxed—I've left the worries outside the door.'

> 'It's the satisfaction of doing something well, and having the machine in your control and doing exactly what you want it to do without any arguments—it's stimulating and exciting.'

Not all of the Dependents used the computer as an escape from human interaction; a great many actually reported an increase in their social experiences which was directly attributable to computing. Not only had many made new friends, but also their expertise with computers was felt to have increased their ability to initiate conversations, with some feeling that their use of the networks had removed many inhibitions and social stigmas.

> 'It's given me contact with others—friendships, happiness in meeting many new friends—I now meet people I like.'

> 'I know a lot more people to talk to and I meet a few as well. I feel more fulfilled and more relaxed now.'

> 'It's broadened my strata of friends out of my social sphere. I now talk to bus drivers, kids, students, company directors, all sorts of people I would never normally meet.'

> 'It makes people more equal—it removes the prejudices against race, age, sex, etc. There's an extra interest to talk about.'

> 'It's made me more sociable and logical. I think I understand people more—they are more honest on the keyboard and say more than they will face-to-face.'

It was of little surprise, in view of these responses, that only one of the interviewees wished to rid himself of his attachment to computing (see Appendix 1, Case Study G). Two others felt that they had now brought their computing under greater control; one because he had found a girlfriend that he did not wish to lose, and the other because he had altered the type of computing that he now undertook—'I'm unhooked from the exploratory stage. It would be arrogant to say I've explored everything, but truthful to say that I've mopped up everything within my grasp. Now I use it for work mainly'.' (However, this man was in fact now spending even more time computing and his wife had noticed no improvement.) The remaining interviewees had no desire to lose their attachment to computing. They gave the impression that it was the most enjoyable activity they had found and could think of no adequate substitute for the satisfaction gained through computing.

The majority of the interviewees did not consider that they spent too much time computing. Many wished that there were more available hours in the day, stating that the need for sleep and the necessity to go to work inhibited them from spending all the time they would have wished. Although recognizing that other activities had had to be sacrificed when computing had become an almost exclusive interest, when asked what they felt they would be doing other than computing, most stated that they would have continued with the hobbies which had been of previous interest—electronics, radios, music, photography, model-making, etc. Some also admitted that they would spend far more time watching television, which most stated they found boring and time-wasting; others would have indulged in more reading; and two thought they would have more time to do the chores which were normally left undone. Only one of the interviewees suggested that he would have spent more time socializing.

One could conclude, reading this extensive list of the positive effects arising from computing, that the Dependents had found the ultimate in hobbies to which most would aspire. As the personalities of the Dependents were able to be so well complemented by their interaction with the computer, many of them failed to understand how any intelligent adult could not become as equally fascinated by new technology to the extent that they were; as expressed by the following interviewee:

> 'For some strange reason that I cannot understand certain people *do not* become addicted to computers, maybe you should interview them, I think they are the strange ones! It's like a box full of answers just waiting for you to ask the right questions. It's like a never-ending jigsaw puzzle. You may keep finding a clue here, another one elsewhere, until eventually all the pieces fit together and you solve your particular problem. With the state-of-the-art technology there is always the thought that with a little more time spent at the computer you may come up with some idea which the boffins have overlooked. With the prospect of such things in mind, the importance of such silly things as painting the kitchen window or fixing

a leak in the conservatory roof pales into insignificance. You have to agree that it is the non-addicts who are the strange ones.'

However, in spite of the Dependents' enthusiasm and the apparent benefits gained from their use of computers, the positive effects remained very personal. Although one could assume that some employers had benefited from the Dependents' increase in knowledge, very few of the Dependents' partners shared their interest and almost none appeared able to recognize any advantages arising from this fascination with computers.

Positive effects upon others

This section is undoubtedly the shortest in the book. Of the letters received and discussions undertaken with partners of the Dependents, only one response could be interpreted as more positively than negatively biased towards computer dependency. The following is an extract from a letter, forwarded to me from the BBC after a radio interview about the research, from a woman whose husband did not in fact take part in the research. (In addition, she also enclosed a story entitled 'An Apple each day keeps the man at play', wittily bemoaning her fate as a 'computer widow'.) This woman had reached a stage in her life when she cherished her independence and seemed delighted that her husband had found a fulfilling hobby.

> 'My husband bought his first computer at the age of 73. He taught himself to use it and a year ago, at the age of 78, he progressed to a more complicated machine. He reckons that it keeps his brain alive and stops the rot of old age. His use of it is mostly programming.
>
> 'Although I do frown upon his staying up sometimes to the small hours of the morning, lost in a world of his own in front of the screen, on the whole I find the computer a god-send. It leaves me free to pursue my hobby without a guilty conscience and keeps him from hanging around me in the kitchen, as I fear many husbands retired from business tend to get under their wives' feet.'

Although some of the young interviewees believed that their parents were encouraging and proud of their skills, the only other direct positive responses came from a few wives of the Dependents who reluctantly suggested that computing was preferable to other activities that their husbands could have indulged in, such as womanizing, drinking or gambling. In addition, some did admit that they were pleased that they knew where their spouses were at night, and a few stated that they did appreciate the increased incomes which had been earned by their partners because of their computering expertise. Other than these rather neutral responses, the Dependents' computing activities tended to be seen as having rather negative effects upon marriage and family life.

Negative effects upon the dependent individuals

Bearing in mind that an overwhelming majority of the Dependents did not wish to rid themselves of their dependency upon computers or considered that they spent too much time computing, some did admit that they experienced negative effects from their activities, although these rarely outweighed the positive benefits gained from their increased satisfaction. Comments such as 'It's clean, it's kind to trees—there are no disadvantages', 'No, there can't be any disadvantages, they are so fantastic' echoed the opinions of the majority.

Very few of the interviewees felt that negative effects had caused their paid employment to suffer. One admitted that he had in the past spent time playing with the computer while he was at work, but stated that this had stopped when he bought his own home microcomputer upon which he had unrestricted access to program as he wished. Another had found it difficult to concentrate at work, and instead found himself puzzling over computer problems which he could only explore while at home. This seemed to be especially problematic when he was cycling home from work, when on more than one occasion he had nearly ridden into parked vehicles. The problems of a third interviewee seemed to have occurred because of a poor match between his skills and those demanded by the job since his last promotion. Because of his expertise with new technology he had been upgraded to managerial level, and he recognized that his management of others was poor as he preferred to spend all of his working hours directly with the computer.

A few of the other Dependents, perhaps more aware of their strengths and weaknesses, had been reluctant to take promotions to managerial levels for these very reasons; they preferred to remain in more lowly posts or take sideways job changes if they could still devote their time to computing. This seemed to apply particularly to those in educational establishments for whom promotion frequently meant greater responsibility for others. For the majority, however, new jobs had provided them with a more satisfactory and satisfying outlet for their talents, to the extent that two of the Dependents now spent far more hours at work as a consequence. This had nevertheless tended to add to the problems experienced by their partners and families with whom they now spent less time.

Those who were still studying, either at school or university, usually did admit to achieving lower standards academically because of their devotion to computing. 'My school work suffers', 'It was an impediment to my degree, I should have been reading books', 'It detracts from my schoolwork', 'I did very poorly at university in the first two years—it was detrimental to my coursework and I nearly got thrown out for hacking into the university's Prime system' were such examples. In spite of these comments, these young men all felt confident that they would still find jobs within the computer industry easy to obtain because of the skills they had gained.

> 'It's the most fun thing I've ever found to do. It's got me lots of job offers—they didn't even want to know about my degree, just my computing experience.'

This appeared to be confirmed by a man with no tertiary level qualifications who had achieved a senior post because of his confidence and experience with new technology. In addition, some of the younger Dependents had already gained prestige and status by having articles published, by interacting with adults on an equal basis, and by working as free-lance programmers during the holidays. They felt that the skills they had were saleable, providing they found posts which utilized their attributes.

Some did admit that their social lives had been adversely affected, as some now spent far less time with old friends most of whom were uninterested in computing and with whom they now had little in common. (Nevertheless, some of these same people had often increased the number of people they referred to as friends from amongst the contacts they made through computing, either face-to-face or over the networks). Others did however use their computing as an escape from social interaction, which in the past they had usually found stressful; now they no longer had to try.

> 'I miss out on social activities, but I don't mind, it's a convenient alibi. I'd rather be computing.'

> 'It's given me something to escape to. If I hadn't had it I would have tried to escape in other ways—lots of walks. I cannot stay in other people's company—I have to be alone.'

> 'I have close friends I used to visit a lot. Since I had the computer I don't bother with them. I don't go to church any more, I don't need it.'

Although many admitted that they did feel cut off from their old friends, this rarely caused them real problems. They accepted their low level of need for people and many were more content when alone and working on the computer. They did not enjoy loosely structured social activities and many seemed delighted that they now had an excuse for not attending such events. Their circle of friends had altered, and tended to be confined to those who were equally interested in computers. However this shift in bias, from one type of friend to another, could not be effected within the home, and the home was the site of the most serious negative effects.

The greatest problems arose from Dependents who were married, many of whom no longer spent much time with their spouses and children, especially watching television with them (which they considered to be an activity undertaken to relieve boredom and not a social activity to be shared). The problems their isolation caused to the members of their families naturally became their own problems by default. Nevertheless, few felt that the problems were entirely theirs, although two were prepared to take at least some responsibility for the situation:

> 'It bothers me inasmuch as it bothers other people—if it bothers my wife it bothers me. I feel guilty that I'm not doing other things.'
>
> 'Yes, it bothers me, but maybe because it bothers my wife.'

One man, who had only been married for five weeks, felt that his computing could lead to domestic disharmony, and another felt that it had adversely affected his sexual relationship with his wife. A few were concerned by the amount of money they spent upon the activity and upon their telephone bills, and some were also concerned that they no longer had time for other activities.

When the interviewees were asked whether they believed that computing had broadened or narrowed their interests, intellect and personalities, although the majority clearly believed it had broadened these aspects, a few had to admit that it had not:

> 'My intellect—no, it's probably narrowed it. I tend to think of things in terms of computing, but I'm more logical and have extra interests now.'
>
> 'I think it has narrowed them really. I hardly ever listen to music or watch TV but I get out a lot with interesting computer people.'
>
> 'Broadened them? Not terribly, in some ways its narrowed them, I've lost interest in other things.'

Most of the interviewees felt that their experience with computers had left their personalities untouched or had allowed their true characters to be expressed more fully, but a few felt they had been adversely affected. Two believed themselves to be more intolerant and impatient with others, but the experiences of a third were more serious (see Appendix 1, Case Study G). John had experienced social problems throughout his life, and because of family problems and his inability to communicate fully with his wife had used the computer as a means to escape from real life. After a trial separation and some treatment he and his wife were now back together. He had learnt how to communicate his feelings far more effectively with his wife, and they now described their marriage as very happy. Although he no longer worked with computers he was well aware that computing was not the cause of his problems, although he realized that it did emphasize and exacerbate them. The computer had offered him control and an increase in self-esteem and confidence, which he has now found through a deeper, sharing relationship with his wife.

Another correspondent also felt that his computing had come to dominate his life adversely. In his late thirties, he was an unemployed, married man with two children. He was one of the few Dependents who expressed any interest in politics and was a local councillor, a position which also took up much of his time, as did his interest in football. The following is an extract from his original letter.

> 'I can confirm that being a code junkie is not a figment of someone's imagination, that the syndrome really does exist. A month ago, my wife laid it down the line—either the computer goes or she will divorce me. That is not an idle threat. Though she has not got grounds on computer desertion, we have argued more in the last six months than our whole 10 years of marriage: my children hardly know me: my personal appearance has deteriorated: I have become estranged from my family in every way. My wife became distant and certainly distanced herself from me because she couldn't understand what was going on.'

Eighteen months later he and his wife had separated, 'so we can sort things out—I'm afraid it is an occupational hazard with councillors'. He believed ultimately that he and his wife were incompatible, and that his computing had merely highlighted their differences.

> 'Funnily enough, computing and football don't seem to figure in the reasons for the family upset, although in retrospect my spending time on both activities took up all my non-council spare time and that *could not* be fair. Nevertheless, separation from the keyboard seems far more traumatic than separation from my wife, although separation from football is just about bearable compared with separation from my children. Odd, isn't it?'

What had started as a pleasurable activity for these two men had become an escape, not only from the problems which were caused by their computing but also from other areas of life, to the point where it had taken over and dominated their waking lives. Neither wife could tolerate the situation and both couples had separated, although in the first case this was temporary and their marriage had survived. At the time of writing, the second correspondent seemed to have opted for his activities and had sacrificed his marriage.

Negative physical effects were few. Although most enjoyed the adrenalin 'highs', John's had become physically unpleasant and one man stated that his dependency led to restlessness, sleeplessness and tension. His general practitioner recommended that he have an affair; an unusual medical prescription one assumes!

The Dependents often went without food for many hours, and most seemed to have adapted to needing very little sleep. 'I only sleep from 4 until 9 am' was not atypical, although some did feel exhausted during the day, with one young student stating that he could often 'be out of commission for 24 hours after a particularly long session'. However, in spite of the long hours that the Dependents spent upon their computers, invariably without regular breaks and often at very poor workstations, report of physical complaints were uncommon. Ergonomists, who in recent years have devoted much of their attention to alleviating the visual, psychological and musculoskeletal stresses of those working with computers, are aware that when one is self-motivated, working upon problems which are personally fufilling and at self-determined work rates, such factors can remove many of the complaints reported by those keying at monitored work rates, often at tasks which are personally meaningless and which lack intrinsic interest.

Of the physical complaints which were reported, one person stated that his epilepsy had worsened and another that his migraines were more severe since working at the computer. However, neither was prepared to cut down the time he spent computing, and the latter had since been prescribed a new drug which allowed him to be able to control the pain of his migraines successfully. Six others experienced regular visual discomfort, and seven frequently had headaches. 'My eyes pack in before my back or my patience—the VDU (visual display unit) is the worst part of computing'. 'At times my eyes feel tired—my eyes want to sleep but my brain keeps going. It can give you headaches, and I feel drained after a long session and need to recover'. Neither the reports of visual fatigue nor headaches were surprising considering the time that was spent in front of the screen, as any intense visual work carried out for many hours at a stretch would probably produce such symptoms.

One young man who reported both headaches and visual discomfort showed me his workstation. He was unable to use his bedroom for his computing as he shared this with his little brother; instead he worked in the built-in wardrobe in his parents' bedroom. Here he worked, often for ten hours at a stretch, in a space about one metre by two, with no ventilation and lit only by a 40W light bulb. It was of some surprise that he ever survived until daylight, and the tolerance of his parents must have been extraordinary!

Three people complained of back and neck pains and one of an 'aching bottom', which was not surprising as most were using non-adjustable dining chairs which were often of an inappropriate height for the keyboard. One single man, who reported having backache, also complained of 'dents in the knees'; a new computer health hazard. His workstation was a coffee table and his seat a sofa; the 'dents' occurred when working at this low level with his elbows balanced on his knees for support. One further physical complaint came from a man who missed the sunshine and fresh air. He felt that manufacturers should produce an outdoor computer in order that he could combine his two loves, but in the meantime he stayed indoors.

The negative effects upon the Dependents caused by their computing were, perhaps not surprisingly, less severe than one might expect; they were after all voluntarily undertaking this activity. Only those for whom serious marital problems had developed to a point of crisis seemed prepared to admit that their computing was partly responsible for this, but even they only saw their computing as symptomatic rather than causal. However, as will be observed in the following section, many of the

Dependents were prepared to admit that their computing did cause problems to others if not to themselves.

Negative effects upon others

Few of the single interviewees reported that problems were caused to other people by their computing. Some of their old friends complained that they never saw them and a few felt that their parents became upset at times, but these problems did not appear to be particularly severe:

> 'I try not to upset my parents, although my mother did threaten to axe it when she could never get on the telephone.'
>
> 'My parents tried to stop me during my examinations, but it doesn't bother them otherwise.'
>
> 'It used to bother my mother until I bought her one, now she enjoys playing games on it.'

The greatest problems were reported to exist within family relationships, although even in these cases most of the Dependents suggested that the problems were not terribly severe:

> 'Yes, there are problems but she's understanding and she knows where I am, but I do neglect her at times.'
>
> 'The kids see less of me because of work, but they also play with the computer at home.'
>
> 'It causes some problems to my wife as I am not doing jobs around the home. She can't see at all what and why I'm doing it.'

Although the Dependents' computing did cause problems in many instances, most relationships appeared not to be seriously at risk. This was probably due to the fact that many of the Dependents had been equally devoted to their past hobbies; their leisure activities had dominated their spare time since they were very young, and their spouses were thought to have grown accustomed to their need for leisure pursuits. However, problems had been caused in the past by their previous activities, and two men actually used the words 'she knew what she was marrying' to justify the time they spent computing.

> 'I've always devoted a lot of time to past hobbies—she knew what she was marrying. I do spend more time on computing though, but I would only be watching TV instead.'
>
> 'I've always been very serious about my hobbies—the theatre, photography and music, and they caused problems. I used to break dates when we were courting, so she knew what she was marrying.'
>
> 'I've always spent all my time on hobbies. Previously I was in the garage building the car, I designed and built it, running for the marathon, making telescopes, joining astronomy clubs—yes, they caused some problems too.'

That hobbies other than computing could create problems within relationships was confirmed by three letters which were prompted by the publicity given to this research. These came from women who had married men who appeared as equally obsessive about their hobbies as the computer dependents. The following are extracts from their letters.

> 'I was "widowed" over twenty years ago, to a harem of motor cycles. He joined classes in motor cycle maintenance, two nights a week, and it has never stopped. Realizing if you can't beat 'em, join 'em, with reluctance (at 38) I enrolled to take driving instructions in an attempt to understand what the fascination was. I acquired my own motor cycle but was still none the wiser what the fascination was for the infernal machines. . . . From the woman's point of view do the wives of the men obsessed with their work fare any better (except in the material sense)? For a vast number of women, with men and their extra curricula activities—what is the choice?'
>
> 'I married one of these people (a Radio Ham) who was an addict, but I had no idea that such people actually existed and thought it was only a hobby. Unfortunately I was proved totally wrong. . . . As to the personality of the beast, the man lives in a world of his own. He cannot communicate—always runs away from any form of conversation, he's frightened of conversation in social terms. The man can be described as afraid to face reality, people, tiresome and wretched situations, lazy, impotent, careless, slovenly—just a plain slob. Frankly, I think that all men look for some kind of 'let out' from relationships with women, albeit too much work, sport mad, drink mad, computer mad, or just plain mad.'
>
> 'His main hobby was radio building and then refining. He built stereo equipment of immensely high standard. I always felt that although he loved classical music, he was more interested in the technicalities of communication than in actually listening to the music. I felt the hobby of radio (and to some extent, fiddling about with the car) was some sort of escape route from the demands of myself and the children for attention. This could have caused problems with our relationship, but I had decided much earlier that I would let him have all the time he wanted for his hobby because his work was so worrying—he needed to escape: but I *did* resent the fact that radio was more interesting than me—so I went back to teaching part-time, in order to be too busy to think. . . . He was very shy. He disliked purely social occasions with no real purpose. He did not read people in the general sense. He was quiet, witty, kind, deep-thinking, tough-minded, objective—he didn't believe in 'meaning of life' discussions.'

These husbands enjoyed technological hobbies which had often been of interest to many of the computer dependents in the past, and also appeared to exhibit personality traits and social difficulties very much in keeping with those of the Dependents. The effects of these hobbies appeared to differ acording to the wives' personalities, and a similar range of reactions was described by the Dependents with reference to their partners; some partners reacted with greater anger, others with some degree of tolerance.

Those who reported few problems within their relationships appeared to have partners who were either interested in computers themselves, had serious hobbies of their own, understood the Dependents' need for solitude, or in fact demanded solitude and space for themselves. Of the following quotations illustrating these points, the first comes from one of the Dependents' wives, the remainder from Dependents:

> 'It's such a waste of energy and creativity, he's doing nothing except playing. But I like going to visit old churches and buildings. We have our own hobbies, he probably thinks that mine is a waste of time.'

> 'My wife sometimes gets annoyed, especially when the computer was in the bedroom and I used to print at three o'clock in the morning. But the time I spend doesn't bother her, she likes to be alone. I've always done a lot alone, and she is independent.'

> 'I had the bug before we met, but she was encouraging. She's a programmer and often she will take part and help me. She wouldn't do it alone, but she does it for the pleasure of my company.'

> 'When I have long sessions on the computer, she has long sessions on her sewing machine.'

> 'He started me off. He thought I should be interested in new technology—he's a gadget person. He's happy for me to do it—it relieves his guilt when he's working. He works all hours, and we are night owls anyway. We go to bed very late, at two or three a.m.'

Some of the partners appeared to recognize their Dependent spouse's need for solitude and intellectual stimulation and were prepared to tolerate the situation, but this level of understanding and selflessness was not observed within all relationships. Although some of the Dependents reported that their partners had often initially been very encouraging of their computing, this tended to disappear as the interest grew:

> 'She was initially very proud and encouraging, especially by my rapid rise in salary. Then she began to compare it with another women; she changed from being supportive to absolute rage.'

> 'He was non-committal before I had the computer, but he was not at all encouraging when I got into it because I was spending every minute on it. The kids got instant meals and everyone complained. I was up to 40 or 50 hours a week.'

A few of the Dependents experienced a great deal of antagonism from their partners, and often seemed bewildered by the problems caused by their computing. Although most admitted that they devoted more time to computing than to their other past activities they could rarely understand the hostility generated towards it. Some of the Dependents did not feel that the problem was theirs and many frequently wished that their partners would become equally engrossed in an activity of their own. Many had attempted to try to teach their partners about computing, but this was often unsuccessful.

> 'She blames it all on the computer. I believe that two people can get on with their own interests; if one party is so absorbed and the other is bored to death, it's their problem. She has no real interests at all. She's now very hostile—jealous.'
>
> 'I do sit here computing when the rest of the family are downstairs watching television. But are they not being unsocial to me? Also, what do they have to show after an evening's viewing, when I might have a new graphics demonstration.'
>
> 'We have had our ups and downs. She has always been very tolerant of me, but I've fulfilled my part of the bargain, I was a good provider. But she is not at all enthusiastic, she won't touch it or have anything to do with it.'
>
> 'It caused problems for a few months. We had a terrible row at one point, and my husband told my brother to take me away or the computer and he meant it. Now I do all his accounts on it, so he accepts it.'

Although most of the marriages seemed stable in spite of the problems caused, the relationships of some of the single Dependents frequently floundered and sometimes ended because of their computing. Their partners were not prepared to tolerate the amount of time that they spent on the computer and felt that they were in competition with it—a competition they did not usually win. Two letters illustrated this point well; the first from a computer dependent, the second from a girlfriend:

> 'During the worst of my obsession, I sometimes found it impossible to stop at all. On one occasion, my girlfriend turned in for the night leaving me typing at the keyboard. When she got up to go to work the next morning at eight, she found me still hunched over the computer. She was offended. "You've never stayed up all night talking to *me*", she pointed out, implying that I found the computer more interesting than her. I scoffed at this, of course, and told her it wasn't the same thing at all. Actually, I was just trying to fend her off, so I'd be left in peace to finish writing the program. But later that day, I gave her comment more serious consideration, and realized she was right: using the computer really *was* more interesting to me than talking to her. This I regarded as a comment on my own weaknesses, rather than a criticism of her. Our relationship ended shortly thereafter!'
>
> 'He started buying computer magazines and talking about how useful it would be to have one, he was using a language I couldn't follow. A microcomputer was bought. He hardly left the computer that weekend. Quite suddenly a certain situation was upon me and I couldn't control it. I was not equipped to fight it either. How do you fight a machine? He came home from work and went straight onto the computer. He came off it long enough to eat dinner, but putting aside his plate it was back fully to the computer until 1 or 2 a.m. I hated this machine, I didn't want to have anything to do with it. I realized there was another alternative and I did leave him. Being alone was better than our relationship, if you could call it that.'

Although relationships between single people were sometimes terminated, of the small number of marriages which were seriously affected most spouses remained committed to their partners in spite of the fact that they were often extremely intolerant of the Dependents' computing habits. The following are extracts from the letters of three wives who had made the initial contact with me because of the problems caused within their relationships and subsequently persuaded their husbands to take part in the research. Although their comments did not seem typical of the effects within most marriages, they are included to illustrate the extent of the problems which could occur.

> 'My husband bought a computer approximately four years ago, supposedly for the whole family, but during the years I have seen the machine slowly taking over my husband and our lives. His need for people doesn't really exist, I would say he was introverted, a bit of a loner and would rather read a book than ever attend a party. Before the computer entered our lives, he was obsessive about various hobbies. He never was the type to just sit in front of the television every night, but had to be constructing elaborate models, etc. We don't, and never really have done things as a family.'
>
> 'He bought the computer, presumably thinking that all the family would benefit. It soon became apparent that the two older children were not going to have much access to it as their father was never away from it, and he has regularly stayed up till 3 or 4 a.m. these past four years. He has even stayed up all night. No amount of coaxing or tears of anger and frustration on my part make him any less inclined to peel his eyes off the VDU, and it is most upsetting when this goes on night after night. It is such a totally anti-social hobby.'

> 'My husband (Case Study D, Appendix 1) has always been interested in machines and electronics during the 19 years we have been together. He nearly bought a 1K DIY computer even before the Sinclair burst upon the scene, so the potential interest was there. He purchased the Sinclair as soon as it came out and devoted all of his and *our* leisure time to getting it going and learning BASIC.
>
> Since those early days it is hard to trace exactly the insidious and malignant way that the computer has intruded into our lives by occupying the emotions, the time, the energy and the priority of G. We can see it around us physically by the hundreds of books which line the bookshelves, by the conversion of one large bedroom into a jungle of monitors, printers, charts, keyboards, wires and more wires and the miscellaneous stationery which support the hobby (he had bought more powerful machines as soon as they became available). The physical reality of this hardware is one thing—infinitely worse is the change in mentality which has marked my husband's obsession. One night, when I tried more than gently to get my husband to pass the time of evening in pleasant company with me, over a meal, a glass of wine, a cup of coffee, I was met with a snarl and a refusal to 'play the babysitter' to me.

My husband was always an isolate within the family, never enjoying what I would term normal family rituals (meals, outings etc.) so this was a very useful cop out for him. Looking back, he was always a remote person, and a poor communicator, someone who hated being told what to do, an unsocial (not unsociable) person who sought isolation. The computer must have been the answer to the dreams of such a personality, and he has used the computer to augment his sense of power, in real terms at work, and in his perception of himself as a person. I have, however, often said, most grudgingly, that it is better for my husband to be obsessed with computers, than with other women or drink. Am I just trying to make the best of a very difficult situation?'

'We were married fourteen years ago when my husband (Case Study E, Appendix 1) was already established as a dedicated academic. During the first year or two I was pretty confident that, however absorbing his work and research, deep down I had first place in his affections. Two years ago he fulfilled a long-held ambition and purchased a microcomputer and printer and triumphantly installed it and himself in his study. Since then there is nothing—literally nothing—which can begin to rival the excitement and fulfillment he gets out of learning, improving and playing with his obsession. He never stops using it until after 3 a.m., and clearly begrudges every hour away from it—apart of course from his academic career which has no doubt benefited greatly by this.

He makes a point of coming down to the drawing room for part of most evenings, where he either talks with great animation about his computer activities, which (a) I do not understand and (b) I do not even wish to become better acquainted with; or he will sit politely, with completely glazed eyes, while I tell him incidents of my life, clearly waiting for the moment when he can feel he has done his duty by me and social conventions, and goes back happily and virtuously to his only love—the micro! How does this affect me? Well, I must confess that I feel on an emotional see-saw; there are times when I am happy, but at other times I am filled with bitter resentment and feel more like a housekeeper than a wife.

There is no danger of our marriage coming unstuck, as I realize that he has a great need for me to be there and cope with the practical side of life, and I have as much respect for his really outstanding genius that I can accept the irritations of living with someone completely unconcerned with social events and the affairs of others.'

Numerous very pertinent issues were raised by these wives, the most important of which was their feeling of being personally usurped by the computer. Although their husbands had previously been very seriously involved with other hobbies, their dedication to computing had been taken to greater extremes. It had filled their waking moments to the point where their wives no longer felt themselves to be of importance to their husbands. Sexual relationships were often damaged; a very hurtful situation to the women who believed that their husbands gained more excitement from the computer than from being with them.

> 'He sometimes worked all night long before he had the computer; sex stimulates his brain, he'd then get up to work. I couldn't bear him coming in at three o'clock in the morining all excited, I resented the feeling that our love life was like a commercial break between two sessions on the computer.'
>
> 'I resented it. He would wake me for sex when he came to bed, sometimes in the early hours. I'd fall asleep and wake up later and he'd be gone again. I felt more lonely when I was with him than when he's away and I'm without him.'
>
> 'Our sex life has altered—he has no time. But if he hasn't the time to spend with me during the day why should I be with him at night? It would be easier to handle another woman than his machine—I could handle that one.'

The last response was from a wife with whom I had had no contact before interviewing her husband. She was especially bitter about the effect his computing had upon her.

> 'I'm very deeply hurt by it. I sometimes wish he would die, if I didn't have the kids. A normal day—he has breakfast and goes to the computer—he comes home, drops his briefcase, puts his head round the living room door, says "hello" and goes to the bedroom. It's just a hello/goodbye relationship now. It's an addiction, not a hobby.'

Her husband's version scarcely differed in detail, only in its interpretation and the feelings engendered.

> 'I feel I'm happily married, but she doesn't. It's my personality, I'm undemanding—she demands a lot of me. She hates the computer—she considers it like a third party, like another woman. I wish, sincerely wish, she would develop an interest. If we could both do something on the computer then I feel our lives would be better. I think it is illogical that she sees it as a threat. It bothers me. She feels I don't spend enough time with her. Now it's progressed to the extent that she goes out to her friends because I'm on the computer, and because she is out I'm on the computer. I think I am being treated unfairly—it could be a joint interest.
>
> I'm a loner, I find it difficult to communicate. I have an easy-going nature, I'm not that demanding. If only she would change her attitude we'd be happier, but I put the blame on her rather than the computer—maybe I'm wrong.'

It was of relevance that this interviewee saw the choice of blame to belong either to his wife or the computer, not to himself, a sentiment expressed by other bemused husbands who failed to understand why they should be held responsible for their wives' reactions. Some appeared to spend their time actively escaping from the demands of their wives in order to carry out their own preferred activities undisturbed, as suggested in the following quotation:

'My interest in folk tales and mythology really suffered in the past, I cannot read because I am constantly interrupted by my wife. With the computer I can retire completely, I have a legitimate excuse, and can put as much effort into that undisturbed. I'm addicted to music but deliberately gave it up. I need peace. It it highly relevant that the computer is in a separate room.'

This quotation was from an interviewee with no children, therefore any problems were borne solely by his wife. Those Dependents with children admitted that they tended to see less of them than previously, although most were very happy to share their computing interests with them. However, it did become apparent that some children had become hostile towards computers, with two wives freely admitting that their own influence had created this disaffection in their chidren. One can only assume that children were adversely affected by the lack of a father figure during the weekends and evenings, although their mothers freely admitted that these fathers had in the past rarely taken part in the nurturing tasks.

A number of other wives had been met during the course of the interviews, most of which had been undertaken in the Dependents' homes. Although many were resentful of the lack of shared time with their husbands, the majority suggested that this had always been a problem within their marriages, and one which they had had to learn to accept. Unwittingly, one wife had found a solution, and her husband was making serious attempts to cut back on his computing time. Her years of argument had had little direct effect upon him, until she started to use the computer networks:

'Now I feel personally threatened by her talking to men at night. I often see what she says and the responses. It has led to telephone calls and letters to her—I still haven't recovered from that. I must have neglected her at the time, but I didn't see any problems—I was here, I wasn't out drinking—I can see that now in hindsight. I seriously think that a lot of other men might well find themselves in the same position—women can use the machine for their own ends.'

Although he had attempted to lessen his own computing, in order no doubt to lessen hers, this illustrates well the differences in needs between many couples. He assumed that all she needed was his physical presence somewhere in the house; but while she desired serious human contact he needed serious, impersonal, intellectual stimulation. The fact that she had achieved her aim while using the computer, a medium he was familiar with, brought home the reality of the situation to him.

The frustration felt by many wives was often intensified by the lack of any extrinsically useful output from their partners' efforts. Other hobbies, even if not appreciated, usually had some type of visible outcome (a model was built, the Hi-Fi could be played, the car ran more efficiently), but lengthy code which was never completed, or sound and graphics which were viewed as crude and simplistic, often seemed pointless to the more practical partners. The impression was even gained that if the Dependents had carried out their activities away from home, the loneliness and the sense of loss felt by the wives would not have been so extreme. They were hurt by being ignored within their own homes.

The differences between the personalities and needs of the male Dependents and their wives often seemed quite incompatible, and one might ask why such marriages had ever taken place; an object-centred person matched with a more person-centred individual. This aspect was not formally investigated, although there were suggestions that each person had initially found comfort in the type of person they had selected. Some of the men had been shown to be heavily reliant upon their partners to the point of quite extreme dependency: a role which both might have found extremely attractive, at least in the initial stages.

> 'I don't like the dependency within marriage, upon me and by me. I'm more dependent on my wife emotionally than I would like.'

> 'He has no respect for anything except work (but not in the financial sense). He's totally dependent on me. I have to manage him completely, his clothes everything. Where he has a role in his job he is perfect, confident, but he's too embarassed to buy bread, he never buys his own clothes, he wouldn't ask for directions or complain. He sees me as the ideal mother, to him really, he likes women to be in the kitchen. I think he is frightened of me.'

> 'His dependence is extreme. I went away with the children for a holiday. He had a whole week in which to sit at the computer without me interfering, or so I thought. Six days passed and I opened the door and there he was. He said the house was too empty and he didn't like being on his own. He has always liked to know I'm here and never likes coming home to an empty house.'

The women suggested that they were well aware that they had found men who had usually had little time for other females; intellectuals who preferred to spend their time alone. Initially both partners were highly flattered by the attention they received from the other. The man because a woman had expressed real interest in him, and the woman because she had won a man who usually feigned disinterest; a victory for both. In order to win the woman, the man had to exhibit the fact that he was interested in the more personal side of relationships, and perhaps in many cases when first in love they really did feel able to share themselves fully with another person. However, this situation sometimes changed once the marriage had occurred:

> 'I always wanted to get married . . . It's a fundamental part of philosophy, when you first get to know them (women) you have to explore them as with computers. There was a period when we first met when I neglected my work to devote more time to her, maybe this was false. There is not

a single story of hers I haven't heard twenty times. I spend more time at it (computing) because it is easier, always there ready, and it's more exciting.'

Often, it seemed, the man saw no more reason to woo; the battle had been won and he could return to the normal, less emotional side of life. A situation which caused some distress to the wives.

'I will never forget the dismay and shock I felt when I realized that once we were married he didn't think he had to take me out, and did so under sufferance—and still does.'

'He talked when he courted me, he now thinks it has all been said. But he doen't talk to anyone else either, even with his colleagues. They are all very impersonal in their relationships.'

A comment from one of the single Dependents confirmed that an effort had to be made in order to maintain the relationship. Whether real need or conventions suggested that they should have a partner is uncertain, but the following comment suggests that at least some of the Dependents really had little need of people, a belief that encompassed those closest to them.

'I feel insecure with women. I don't really know why my girlfriend likes me—I feel insecure about it—I feel she'll realize one day. I don't find her or any one person that interesting, but I'm happier with her than without her. I do try and make an effort.'

Why were such men attractive? Not only were some of the women flattered by the attentions of the normally aloof, studious man, but they also appeared to find great security in marrying a home-loving man whose exclusive need was for them alone. Previous experiences in some of the wives' own backgrounds also indicated that such a man was desirable. Two women had been married previously; one to a 'sociable' type who spent every evening drinking with his friends, leaving her alone with the baby; the other to a man who was impotent. Another had to fight his mother to gain him, and perhaps enjoyed the challenge of winning a man whom she said 'disliked women'. Yet another had been deserted by her own mother when a baby and made to feel totally responsible for her father's welfare. She married in order to escape the drudgery and oppression, and initially found freedom in the partnership.

One cannot suggest that such situations were typical, but they do indicate that on some occasions both partners found characteristics in each other which were extremely appealing during the initial stages of the relationship; both partners were considered atypical of their sex with whom some had experienced previous difficulties. These needs, which in part must still linger, no doubt account for the fact that most of the Dependents' marriages were not ultimately at risk because of computing.

The most serious problems which had been created by the Dependents' computing activities were therefore shown to be confined within relationships. The single people may have lost partners, but this did not on the whole seem to be greatly upsetting to them. Only one of the three married female interviewees stated that her computing had caused serious problems at home, although these seemed mainly to be due to her lack of time to undertake her housewifely duties rather than anything more personal. Although the wives of some of the male Dependents also reported that the gardening, decorating and other chores were now no longer performed by their husbands, these problems paled into insignificance in the light of their more personal problems.

Most of the Dependents had been made aware that their computing was causing problems to others around them, and the few personal problems caused to them were usually connected to the guilt and frustration this engendered. Many of the wives considered the computer to be a real threat to their relationships, although few were prepared to act upon this, and reported that they felt totally unable to compete with an inanimate object for their husbands' time and affection.

Discussion

Positive effects upon the individual

The results showed that the Dependents had achieved some very positive benefits from their computing activities, which for some had been able to compensate for a lack in other areas of their lives. As has been documented in Chapter 8, the majority of the Dependents had experienced great difficulties with social interactions from an early age and often felt themselves to be outcasts to some extent because of their interests and personalities. The use of the computer had since given them the opportunity to reproduce an interaction which mirrored their own modes of thinking, and in addition many felt that they were able to communicate more easily with others. The use of the networks enabled them to interact with people of like mind with whom they were able to share mutual interests, and such activity was felt to have expanded their range of friendships.

Their knowledge of computers had improved their employment prospects and their earnings; they felt more useful to others; had increased their prestige, confidence and self-esteem; they had gained fulfilment and a sense of achievement; and had found a relief from stress. In addition they had found an activity which was educational, intellectually stimulating, challenging and creative. One would assume that most people would consider that any activity which could provide all of these benefits could only prove advantageous to the individual concerned.

Leisure activities are often undertaken in order to provide satisfactions not found in the working environment (as discussed in Chapter 7), however the majority of the Dependents had achieved the type of employment which mirrored their hobbies, compounding their satisfactions. The Steering Group on Job Satisfaction (Work Research Unit, undated), expanding upon the findings of Maslow (1970) and Hertzberg (1966), isolated the factors which have been shown to make work satisfying; factors which equally well apply to activities undertaken voluntarily. These include the need for challenge, variety and responsibility; the opportunity for using skills, for social contact and participation; and scope for learning and personal development; all of which the Dependents had found in abundance from their computing at work and play.

Reports of educational, social and psychological benefits have been documented by others investigating the general effects of computing upon the individual, although not necessarily with respect to computer dependency. Papers which concentrate upon the educational advantages usually refer to school children, although their theories apply equally well to those of any age.

Believing that computers are 'machines to think with, not things which think' and because of their non-human characteristics, Pateman (1981) believed that computers have many advantages when used for educational purposes by allowing learning to take place more easily and positively. He believed that the use of the computer eliminated the 'unnecessary obstacles to thinking, creativity, invention, discovery and learning' which tend to occur frequently with traditional methods of learning, where criticism may stifle the learner's enthusiasm. Computer learning takes place at the user's pace, matching abilities and current levels of knowledge, and because of the lack of inhibition and judgment present within the interaction computers are sometimes found to be easier to relate to than people. Computers do not 'criticize' the person, merely the program they have written, thereby allowing self-criticism and development to take place.

Papert (1972) also saw proramming as a liberation from the tyranny which dominates many classrooms. He felt that the non-judgmental responses received when working at a computer would allow people to accept failure in a positive manner, while encouraging them to strive for other solutions. He also considered de-bugging, one of the most attractive parts of computing for many Dependents, to act as a very effective technique for self-education. Unlike normal educational practice, where mistakes are often only corrected by the teacher and rarely referred to again, the act of de-bugging allows the learner to re-evaluate the problem, thereby gaining a more complete understanding of the processes involved. In addition, he believed that computer users would begin to understand the thinking process itself, leading to 'a new degree of intellectual sophistication', as their logical capabilities improved. Although Papert was describing the use of his own programming language *LOGO*, known to only a few of the Dependents, his suggestion that users would learn to extend their knowledge and mastery of the world, increase their self-confidence and improve their self-image, appeared to have been generally effective for the Dependents.

Others, also interested in the intellectual and educational value of using computers, believed that the use of the computer could lead to greater problem-solving skills which could be applied elsewhere in life; as suggested by Eisele (1981), 'simply modeling the analytical processing of computers will result in greater awareness of the method of analytical problem-solving itself'. Furthermore, computer learning is liberating in other ways. Rather than limiting oneself to the known and the feasible, aspects which control much of current educational methods, Levy (1984) recognized that the science fiction aspect of computing led users, such as the Dependents, to think 'the impossible' and then strive to attain it. Similarly, Turkle (1984) believed that personal programming could lead children to develop highly individualized styles of learning, rather than all being expected to conform to the same educational mould, allowing them to develop their own personal strengths, cognitive styles and personalities in a controlled but free manner.

The private but interactive medium of the computer encourages experimentation, and has been found to increase many types of learning where disinterest had previously prevailed. Greenfield (1984) stressed the advantages of word-processors for the encouragement of creative writing, and reinforcing the educational benefit saw a reduction in the hours spent watching television by computer users to be a positive step, with programming allowing them actively to create, rather than accept passively, the output from technology.

Networking was reported to expand the Dependents' circle of friends. Work by Kiesler *et al.* (1987) demonstrated that networkers were found to be far less inhibited when using this method of communication than when face-to-face, and that more equality of participation was observed with less likelihood of one person being totally dominant in group communication. Although no reference could be found to demonstrate that the high level of egalitarianism and lack of inhibition commonly observed in networking communication is transferred to other areas of life, one might suppose that it would be of long-term benefit to those who are socially inhibited. Boden (1981) recognized this apsect of networking in her observation that 'the sense of community and instant accessibility that is fostered by membership of a shared community environment could be liberating indeed'. Surely any increase in communication, even if anonymous and impersonal, will give the shy person a greater understanding of other people and the interactive process, to the benefit of all concerned.

The benefits afforded by the anonymity of computer communication are not solely limited to direct interaction. The use of a program to conduct initial medical interviews was found by many patients to be preferable to that with a doctor (Card *et al.*, 1974). The computer was viewed as more friendly, relaxing, polite and comprehensible. In a situation where communication is frequently poor and seemingly impersonal, the use of the networks and computerized interactions could be extremely beneficial in aiding accurate communication, especially for those such as the Dependents who are unaccustomed to talking about themselves.

Even the use of computer games has been found to aid the development of intellectual growth and social interaction, whether used educationally or as entertainment (see Loftus and Loftus, 1983). Egli and Meyers (1984) found that arcades acted a social centres and places for the development of friendship, in a similar manner to the Dependents who made contact and friendships with others in order to share and copy games software. Because one is challenging the software, computer games-playing was found to decrease competitiveness and increase cooperation between people (Bowman, 1982) as well as provide the ideal environment for the development of competence, self-determination and status. Turkle (1984) recognized the particular fascination with adventure games prevalent amongst the Dependents; 'the Dungeons and Dragons way of thinking, with its thick book of rules, seems more exciting and more challenging than history or real life or fantasy play where the rules are less clear'. Within its framework, logic and 'lateral thinking' lead to challenging new modes of exploration.

Malone (1984) concluded that there were distinct and contradictory differences between the ideal characteristics of tools and toys. He believed that good games should be increasingly challenging and somewhat difficult to use; not attributes desirable for tools. As most of the Dependents viewed the computer primarily as a toy, whether playing games or programming, they were able to rise to the challenges naturally presented to them by new technology, thus increasing their enjoyment of learning as new problems arose. Malone reiterated traditional learning theories by recognizing that the computer's immediate feedback, positive reinforcement, and the user's active involvement all encouraged learning; suggestions which differed little from Skinner's advocacy for the 'teaching machine' (1954). Malone found that the multiple level goals, the intrinsic stimulation, and the uncertain outcome of many games (similar to the problems encountered when programming) could lead to an increase in the pleasure of learning, as was reported by the Dependents.

In addition, Malone believed that the playing of games could have positive psychological effects through the increase in self-esteem which he frequently observed amongst the players. Similarly, Nicholson (personal communication) believed the effect of having total control over the computer could lead to an increase in confidence which could be transferred to other areas of life. Stating that lack of control is one of the major symptoms of clinical depression, he believed that the use of computers 'may in time come to play a valuable part in the treatment of seriously depressed patients'. This echoed the responses from members of the Tavistock Institute, London, to whom I was invited to give a presentation about this research. The psychologists and psychiatrists present were aware of factors which can add satisfaction to life while removing some of its frustrations. Having heard details of the Dependents' personalities and their difficulties with social interactions, a member of the Institute asked whether I thought that personal computers should be prescribed as therapy to those who experienced stress in these areas of life. Although this was offered somewhat in jest, further discussion revealed that the members present seemed well to understand the benefits which could be obtained from this type of interaction.

In spite of the multitude of positive effects observed, or recorded by others, the overwhelming conclusions drawn by most authors nevertheless were that an exclusive interest in computing could be damaging to the individual concerned.

Negative effects upon the individual

The few negative effects reported by the Dependents included their lack of contact with old friends, although this seemed often to have been compensated for by the creation of new friendships or by the recognition that some really had little need of people and personal interaction. The personal negative effects of most concern were reported by the young Dependents still in education, who admitted that their studies often suffered from lack of attention because of the time spent computing. However, even in these cases most of their parents seemed to have encouraged the activity, except when examinations loomed, and most of the individuals felt that their computing expertise could only be advantageous for future employment. One could therefore conclude that as far as the individuals were concerned any negative consequences were greatly outweighed by the manifold positive effects from computing. The Dependents appeared to have achieved in their working lives and through their hobbies a level of intellectual stimulation and satisfaction to which most would aspire.

In addition, in spite of the long hours they spent at the keyboard, few reported any of the physical complaints commonly associated with working with new technology (see Pearce, 1984), a fact which should be noted by those designing jobs for people. If the task is intrinsically interesting and if the individual is personally motivated, interest can be maintained for many hours without complaints of negative side effects; a theory which many ergonomists are actively teaching and pursuing.

Effects upon others

The positive effects upon others arising from computer dependency were minimal, and the results showed that the most serious negative effects occurred within the Dependents' marital relationships. Nevertheless, fewer problems were reported in marriages where wives recognized their Dependent partner's needs for solitude and intellectual stimulation. Positive, tolerant action on the part of the woman seemed to be the only thing which allowed the situation to stabilize to a point which was acceptable to both. The vast majority of the Dependents had been heavily involved with hobbies other than computing and many partners were familiar with this situation, as has been acknowledged by others.

> 'If the husband had not been making love to his computer he might have been sleeping with a vintage car, and who's to say which is the worse?'
> Boden (1981)

Many people, Dependent or not, experience marital difficulties and appear to find the members of the other sex somewhat incomprehensible, but this does not seem to prevent them from forming and maintaining inter-relationships in spite of difficulties which may develop (Hite, 1988). Many also do not seem to desire a marriage of like-minds, but a complementary personality to their own. It seems that only after some time in marriage do these differences become problematic. Problems seem to arise after the first flush of love has died away, when people revert to their old familiar patterns of behaviour and their true personalities flourish once more. In the case of the Dependents' relationships, only then did the women discover that their men were not going to become gregarious and share their feelings and emotions, any more than the men were able to interest their intelligent women in the abstract joys of computer programming. Both became disappointed, but it is quite likely that this would have occurred to an extent even without the computer.

Only a small proportion of the Dependents or their spouses reported problems of any severity caused by computing, and none of the interviewees had attempted to gain any professional held for their marital problems. It seems that Americans are perhaps less inhibited than the British about discussing personal, social and psychological issues, and Jean Hollands, a psychotherapist and marriage counsellor in California, now devotes almost all of her time to couples in similar situations as the Dependents and their partners. Working in Silicon Valley, she realized that special problems arose when 'a scientifically-oriented thinker mates, works with, or loves a more emotionally-oriented person: the union of an engineer/scientist with a non-scientific partner . . . the "thinker" with the "feeler"', (Hollands, 1985). Although most of the Californian men (once again it was invariably the male who was the scientifically-oriented partner) could be seen as 'workaholics' and 'company men' rather than computer dependent, their marital problems seemed similar to those described in this chapter.

In her self-help book, Hollands (1985) gives valuable advice for couples, although even she seems to suggest that it is the wives who need to do the adapting and understanding and take charge of the sensitizing of their husbands. There is a chapter entitled 'How to love your engineer/scientist' but no reciprocal one about how to love your non-scientist. It seems that because the woman tends to be the sufferer in such situations, and perhaps because they are the 'emotional housekeepers' (Hite, 1988), the onus rests with them to affect the changes, without which the stalemate continues. Hollands discovered that even if the husband was unhappy with the resultant difficulties, this type of man was invariably ill-equipped to know how to understand the woman's position without some form of positive emotional re-education.

> 'Some men are already inclined to withdraw when they feel the pressure of difficulties in a relationship: the computer encourages this withdrawal—so making the problem worse.'
>
> Simons (1985)

As suggested by Simons, problems had not arisen in the Dependents' marriages where none had existed before, but difficulties within a few couples had increased considerably since computing had become of interest to one of them. From the results, one may assume that as far as the Dependents were concerned, little had changed. Their personalities had not altered, but their preference to spend time alone with the computer had adversely affected some relationships. The Dependents had not altered their attitudes towards people in general or their partners in particular, but had found what one might call the 'ultimate interaction'; one which was difficult to share except with another of like mind. They had always found activities which had been able to give them control and intellectual stimulation, and in some ways the computer was little different except that it offered more of the positive effects than had the others.

Many of the Dependents had been unhappy for most of their lives and had now found satisfaction, sometimes at the expense of their relationships. A few wives quite justifiably felt reduced to the level of housekeepers and had lost much of their own stability and *raison d'être* by the arrival of the computer in their home. Marriages of different rather than similar personalities are perhaps always more likely to be fraught with problems, and stalemate would no doubt continue until the Dependent partner lost interest in computers, which seemed an unlikely event, or until one or both realized that some positive action was needed. Unfortunately, but not atypically it seemed that this action had to be undertaken by the female, and that she would possibly need some external assistance in this task.

Conclusion: weighing the balance

The results demonstrated that for most of the Dependents computing proved to be personally advantageous, by increasing their self-esteem and confidence, removing depression and adding to their measure of happiness. The Dependents were shown to be people who desired constant intellectual stimulation and for them this could be more easily gratified through interaction with an inanimate object than with other people. Computers were not seen to have *created* problems within relationships where none had previously existed, but in some cases had emphasized and exacerbated existing difficulties.

As more of the Dependents acquire modems, some may find that their circle of friends steadily increases as the computer networks become more prolific, and confidence within these types of interaction may be extended to relationships carried out face-to-face. Perhaps, more importantly, it may also lead to a greater awareness on the part of the Dependents who felt isolated and alien within society that there are others who possess similar interests to their own with whom they can learn to share and whom they can trust.

The shy and introverted child, if encouraged to share his knowledge about computers, could increase his self-confidence and extend his interest to other intellectual activities, and the feelings of control engendered by computing may also prove transferable to other areas of life. As suggested by Simpson (1983), computer dependency is a sign of deficiencies elsewhere in life, and unless more enriching experiences are provided which match the needs of the Dependents, it may be in their best interest that they have found such a satisfying activity with which to fill their time.

Chapter 11
Discussion of the findings: Should computer dependency be considered a serious problem?

Introduction

This research was undertaken in order to determine whether and why computer dependency occurs, to examine its effects, and to ascertain whether the fears and anxieties shown towards this syndrome have any foundation.

The results from this exploratory study revealed that although computer dependency did occur for a small proportion of computer users, the effects arising from it were not as dire as suggested in the literature. Furthermore there were logical reasons why some people chose to turn to an interaction with inanimate artefacts for the satisfaction of their needs. What was less clear was why the cognoscenti should have voiced their fears so forcefully, and why alarm was expressed for those who used computers intensively when dedication to other activities was often encouraged.

The discussion within the first part of this chapter puts forward the theory that computer dependency was only one of a multitude of computer-related anxieties current at the time this research was initiated, many of which had been well-publicized in the press although not all were necessarily well-substantiated. It is posited therefore that the fear that some people would become socially and psychologically dependent upon computers was merely one of many fears attributed to computer usage and was not atypical of what one might expect. The introduction of this new technology appeared to have created a climate of apprehension about computers in general, and any overt behaviour patterns associated with their use were perhaps seen as suspect.

That these worries occurred was not in dispute, but they were found to differ little from the anxieties which surrounded the introduction of many other technologies over the centuries. This is not to suggest that such misgivings should be ignored or dismissed, but they have been shown to occur with a certain regularity when people do not fully understand the ramifications of a new technology. The discussion in the following section suggests that anxieties about technology have been ever-present, and that time and greater understanding and experience tend to allay them somewhat as the artefacts themselves become commonplace.

Much of the remainder of the chapter deals with specific suspected concerns related to computer dependency, including effects upon the working environment, upon children and upon the wider population.

An attempt is made to deduce why many authors had believed that extensive computer use could be damaging to the individual concerned and why computer dependency had been singled out for their attention. It is suggested that intensive interest in computers differs little from dedication to many other activities. 'Dependency' upon one particular pastime was not shown to be unusual, and other devotees were found to show very similar characteristics to those observed within the computer dependent population; the difference appeared to lie in the attitudes of others towards what were considered to be acceptable pursuits.

Anxieties associated with new technology

The introduction of computers into our society occurred with great rapidity with the development and widespread use of the microchip, and was often poorly executed without adequate information being given to those who were expected to use them. This lack of knowledge and understanding appeared to some extent to have been influential in the development of many and diverse anxieties being associated with the use of new technology. Many within the population were extremely sceptical, if not specifically fearful, of computers; therefore any people who showed a serious interest in new technology were viewed as atypical and perhaps somewhat suspect, especially if they were the type who were generally not well-understood by others. It is suggested that the anxieties associated with computer dependency and the effects arising from it may be no more unusual than many others which have been attributed to new technology.

The 'computer revolution', although developing slowly during the last few decades, exploded in Great Britain in 1980 with the introduction of the first mass-produced microcomputer, the *Sinclair ZX80*. Until that time the computer had remained within the domain of industry, science, academia and the serious electronics hobbyist; now it was available to all, at a price less than that of a colour television. The domestic market accelerated at an astounding rate as did the installation of terminals and microcomputers in the workplace, but along with the computer came many fears and suspicion.

The introduction of any new technology may be considered cyclical; from the development of the science and its application, to the mass production and widespread dissemination of the machinery. This appears to be followed quickly by concern about the possible negative results flowing from its use; concern about the environment, society and the individual. However, most older technologies only directly affected a few people at their inception, and most systems were well accepted for their benefits before they became the common property of society. This appears not to be so with computers; their introduction was extremely rapid and seemed to affect the lives of those who have never directly used or even seen them. They were held responsible for faulty wage packets and invoices, endless delays, and even serious accidents. Despite the thought that the computer was 'intelligent' it was also believed to be uncontrollable and to have no conscience or morals, as humans seemed unwilling to take responsibility for any faults in its output.

Computers affect most western people, thousands of whom have to work with them, many of whom remain suspicious. Quite regularly the press bring computer-centred scares to our notice and few can be thoroughly disproved to the satisfaction of society. Each nation (and one may probably link this to the influence of the press) seems to develop its own local computer 'health hazard'. In Australia computer users seem to be particularly susceptible to the crippling physical condition of tenosynovitis (Ferguson, 1987), in the USA cataracts are thought to be caused by the radiation emitted by VDU screens (Zarat, 1984), the Scandinavians suffer from facial rashes (Tjonn, 1984) whilst Europeans seem more concerned with postural effects (Grandjean and Vigliani, 1980).

In Great Britain various documents, especially from within trades unions, echo these fears by suggesting that the use of computers could be responsible for such physical problems as visual discomfort, ocular damage, postural fatigue, tenosynovitis, facial rashes, epilepsy, and even miscarriages and birth defects. (See the *ASTMS Guide to Health Hazards of Visual Display Units*, 1979; the APEX document, *New Technology: A Health and Safety Report*, 1985; the *TUC Guidelines on VDUs*, 1986; and the *NALGO Health and Safety Briefing*, 1986.)

Fears often develop through suspicion and lack of knowledge, and although some of these anxieties may be ill-founded the response from ergonomists confirms the fact that these claims are being taken seriously and are under rigorous investigation. For a review of these issues see Berqvist (1984), Cakir *et al.*, (1980), Grandjean (1984) and Ong *et al.* (1988).

The new, unknown technology is frequently seen as more threatening than older mechanisms which have been proven to be dangerous but are now familiar. Ergonomists, consulted in order to reduce the 'health hazards' associated with computerization, often encounter dichotomous situations.

> 'I have walked through factories full of noise, dust, and noxious fumes, past unsecured ladders and dangerous equipment to reach an office in a corner. Here it is relatively quiet, and has a pleasant thermal environment where workers sit at comfortable desks using keyboards. I have been called

into these offices because they are worried about the health hazards of VDUs.'

Stewart (1984)

This quotation sums up much of the apparent irrationality of the fears expressed towards the new technology. Our lives are normally surrounded by risks and hazards. The high death toll on our roads will not lead to a governmental ban on motor vehicles nor to the majority of people becoming fearful of driving their cars. Although the medical hazards of smoking are well documented, many continue to smoke and new smokers perpetuate the habit. The risks are clear and obvious, we make choices from the available information and take the gamble that we will be safe. The same applies within factories and upon farms and building sites; the hazards are well understood and obvious. Acts of Parliament have made industrial and agricultural work much safer, but still there are deaths. Productivity is the important issue, and further precautions are not taken because of the financial and political implications which would occur through loss of revenue.

However, unlike the machinery of heavy industry, the risks of working with computers are neither obvious nor well-understood. Manufacturers assure us that the radiation levels are well within the safety limits, but the setting of these limits could be seen to be arbitrary and is of little comfort to the pregnant VDU user who is worried about the health of her child. One cannot see radiation, as one can the traffic on the road or the pool of oil on the factory floor, and the mystery of the computer remains hidden from the majority. To those for whom it does not, they themselves become mysterious and incomprehensible and seem unable to communicate in laymen's terms in order to allay the public's fears.

Any potential psychological effects are less tangible, and therefore more disturbing than the more obvious accusations of physiological damage caused by computerization. Anxieties and stress-related reactions towards working with new technology have already led to the coining of the term 'cyberphobia' by Weinberg (1982); a condition he describes as exhibiting the characteristics of rapid heart beat, nausea, diarrohea, sweating, and so on. Generalized fears seem to concentrate upon the belief that new technology strips people of their freedom and privacy. Thimbleby (1979) saw the potential for irreversible totalitarianism and oppression caused by the concentration of power in the hands of the few, and cites the example of police databases. Many believe that the computer professionals, whose programs are able to create such situations, are ill-equipped to perform their functions adequately. They are seen as lacking humanitarian concern, centering solely upon the capabilities of the computer rather than on the societal effects (Rothery, 1971).

Computers have been accused of causing such psychological and sociological effects as the de-skilling of labour (Tuckman, 1984), job stress, redundancy and unemployment, social isolation, powerlessness

(Mendelson, 1983) and alienation (Sprandel, 1982). Further, the use of computers has been charged with altering personalities (Weinberg, 1971), with changing the socially gregarious into recluses and destroying relationships, and Simons (1985) even speculates that if:

> '. . . the computer proves to be a quite sufficient companion, then the seeds of anthrophobia (fear of other human beings) may develop in the obsessive programmer, the hacker . . . the computer is clearly a potentially fertile source of phobic conditions.'

Papert (1980) also saw the possibility that computers could damage human relationships and create social differentiation between the cognoscenti and the 'computer illiterates', exacerbating the existing class distinctions rather than reducing them. Although personally believing in an optimistic computerized future he also believed that use of the computer could alter the way people think, reason and react:

> 'If the medium is an interactive system that takes in words and speaks back like a person, it is easy to get the message that machines are like people and that people are like machines. What this might do to the development of values and self-image in growing children is hard to assess. But it is not hard to see reasons for worry.'

Weizenbaum (1976) even described new technology and some of its applications as potentially 'obscene', 'morally repugnant' and 'dangerous', and attacked the dehumanizing aspect of computerization and its impact upon society. 'I'm coming close to believing that the computer is inherently anti-human—an invention of the devil.' One might believe that he is attributing anthropomorphic, super-human characteristics to a mere electronic tool. If such mistrust and apprehension occur to those who have spent years working closely with computers, who understand them and can 'control' them, what chance have the rest of the population to believe in the worth and goodness of the machine?

The fears cited appear not unrelated to the new anxiety, that certain people seem to become 'addicted' to their interaction with the computer. Especially when children are affected in this way, the machine appears to be seen as a potential source of danger and subversion. Children seem to be less suspicious of new technology than older generations, with many quickly learning the jargon of the professionals and showing little reluctance to use computers. This in itself can cause problems. Parents become concerned when their children appear to cut themselves off from social activities in order to 'play' with the computer for hours at a time.

This differs little from the fears expressed thirty years ago with the advent of another revolutionary technology into the home, the television. The television was accused of affecting children in very much the same way that the computer is today, to the extent that various research programmes were established to investigate its influence over young minds. As suggested by Weizenbaum (1976) with respect to computer games, the common fear expressed with particular reference to children and television was that they were being adversely influenced and affected by violence on the screen. It was thought that children could be directly influenced, to the extent that they would imitate the actions seen, and become inured and habituated to it. Although disputed by Halloran (1964), research by McIntyre and Teevan (1971) and the Report on the Future of Broadcasting (1977) (known as the Annan Report) seemed to confirm these beliefs, and the potentially damaging influence of television is still being debated (see Howitt and Cumberbatch, 1975 and Cullingford, 1984).

A more pervading fear lay with the fact that many people seemed to be cutting themselves off from other forms of recreation by watching television excessively, and once again children were thought to be especially susceptible. Himmelweit *et al.* (1958), undertook a special study to investigate the young 'television addict' and hypothesized that:

> 'Emotional insecurity or inadequate facilities at home cause the child to become a heavy viewer. In this way he restricts his outside contacts and so reduces still further his opportunities to mix. With escape through television so readily available, other sources of companionship may demand too much effort and offer too little promise of success.'

Although some people still remain concerned about the levels of violence and even sex and 'bad language' on television, the general population no longer seems to consider television watching to be overtly damaging. Television viewing has become one of the most popular forms of recreation in Britain, and a recent survey showed that the 'average viewer' watched 27·8 hours of television per week (Miller, 1988). This is less than the average time spent computing at home by the Dependents in this study of 22·4 hours per week. One would assume that any activity, which could provide the benefits and advantages listed by the Dependents in Chapter 10 as flowing from their interactive computing experiences, would be seen as preferable to somnolent evenings spent in front of the television set. However, few have recognized these advantages (see Greenfield, 1984 and Nicholson, 1984) and most remain suspicious; even those who have professional knowledge of computer systems, such as Boden (1977):

> 'The socially isolating influence often attributed to television is as nothing to the alientation and loneliness that might result from over-enthusiastic reliance on the home terminal and associated gadgetry.'

Although a computer scientist, or perhaps because of it, Weizenbaum (1984) expressed more doubts and fears than most and was very concerned about the influence of science in general, and computers in particular:

> 'All thinking, dreaming, feeling, indeed all outer sources of insight have already been deligitimated. The indoctrination of our children's minds with

> simplistic and uninformed computer idolatry, and that is almost certainly what most of computer instruction is and will be, is a pandemic phenomenon.'

Although apparently extreme, the fears directed towards the computer were not entirely new and differed little from those expressed towards television. Further research was able to show that such negative reactions had been expressed for centuries with reference to the introduction of other older technologies, and for this reason such reactions could have been anticipated.

Fear of technological innovation is not a new occurrence

> 'Were we required to characterise this age of ours by any single epithet, we should be tempted to call it, not an Heroical, Devotional, Philosophical, or Moral Age, but, above all others, the Mechanical Age. It is the Age of Machinery, in every outward and inward sense of that word; the age which, with its whole undivided might, forwards, teaches and practises the great art of adapting means to ends. Nothing is now done directly, or by hand; all is by rule and calculated contrivance. For the simplest operation some helps and accompaniments, some cunning abbreviating process is in readiness. Our old modes of exertion are all discredited, and thrown aside . . . For all earthly, and for some unearthly purposes, we have machines and mechanic furtherances.'
>
> Thomas Carlyle (1829)

Over one hundred and fifty years ago Carlyle prophesied the corruption of his civilization, with sentiments which scarcely differ from those observed today. As has been suggested with reference to computer use, he also believed that the psyche as well as the physical world could be altered by the introduction of new technology.

> 'Men are grown mechanical in head and in heart, as well as in hand. They have lost faith in individual endeavour, and in natural force, of any kind. Not for internal perfections, but for external combinations and arrangements, or institutions, constitutions—for Mechanism of one sort or other, do they hope and struggle. Their whole efforts, attachments, opinions, turn on mechanism, and are of a mechanical character . . . Not the external and physical alone is now managed by machinery, but the internal and spiritual also.'

The working population also expressed their displeasure towards new technologies, sometimes in the form of physical rebellion. The legendary figure of 'General Ned Ludd' from Nottinghamshire (who achieved notoriety, it is said, by leading men to smash the stocking frames whose introduction had led to the lowering of wages) lives on today. However, the stocking frames survived and society as a whole was not destroyed by the advent of the industrial revolution, although there is little doubt that lives were irrevocably changed. Only a few decades ago the majority of the British workforce was involved in agriculture, now only a few per cent of the population are needed to supply our food needs; a fact which would have astounded our forefathers. Man does adapt to the new roles prescribed by technology, although the term 'Luddite' is still used to describe those who question the wisdom of the widespread introduction of mechanization and remain hostile and suspicious towards it.

There will always be opposing points of view with the advent of any new technology, as in most cases there are no right or wrong answers but merely different interpretations of the beliefs which are current. It is impossible to dispute that inventions such as the steam engine, motor car, wireless, telephone and television have altered civilization quite substantially, many might say detrimentally; but there are few in the western world who would forsake the benefits and freedoms gained from their introduction. However, at some time or another each has been accused of damaging those who come into contact with them, either psychologically or physically. Gas lighting was almost universally considered dangerous and even impossible by many. However its social benefits were quickly accepted after its inception, subsuming the more irrational fears. Russell (1976) stated that winter schooling was extended and street crime dropped dramatically; factors both directly attributed to the implementation of gas lamps. Even the introduction of the telephone led to a great deal of resistance, the common fears being that its use could lead to deafness or electrocution. However, Vallee (1984) has more recently concluded that most of the apprehension and resistance was due to the new mental barrier which was created between the speakers.

Thus many apochryphal tales abound, but most are difficult to substantiate. At one time travelling at more than 20 miles an hour in a train was thought to make the body disintegrate. When people survived at this speed, travelling at more than 40 miles per hour was said to lead to suffocation as it was thought that all the air would be sucked from the carriage. Air travel was thought ridiculous, 'Flight by machines heavier than air is unpractical and insignificant, if not utterly impossible' (Simon Newcomb, 1835–1909, cited by Pile, 1980). When this was disproven by the Wright brothers, it was later thought impossible that aeroplanes could travel at more than the speed of sound, and the effects of long-term space travel still remain something of a mystery.

Mistrust has always existed and has always to find an outlet. These examples hardly differ from the present fears that working at a computer can lead to ocular damage, miscarriages or computer dependency; anxieties are voiced and suspicions are aroused, and they are rarely disproved to the full satisfaction of society.

> 'Processes which have gone on quite steadily for centuries may have to stop quite suddenly and quite soon . . . Many of our traditional ideas have

> simultaneously and abruptly become obsolete, and we do not know how to adapt ourselves to the new situation.'
>
> Bowden (1965)

This opinion from Lord Bowden, then Minister of State for Education and Science, came in anticipation of the impending changes which were to occur with mass computerization. A few years later, Toffler (1971) published '*Future Shock*', his treatise on the 'disease of change', which greatly expanded upon this concept.

Anxieties such as these have not always been assuaged by the scientists themselves:

> 'It seems to me that our only chance for the future is to recognize ourselves for what we are—the priests of a not very popular religion—and to see what we can do about it. If society will not accept us for what we are, we must change society.'
>
> Hoyle (1968)

Society has changed, but not in a manner to endear itself to science or scientists. Especially since the dropping of the first atomic bomb attitudes have, as suggested by an historian, altered to the extent that science in general is now mistrusted:

> 'Science is no longer a joke. It is more often seen by non-scientists as something dreadful which is profoundly damaging to the human race, either by threatening to destroy it or by transforming its way of life.'
>
> Duncan (1986)

There may be good grounds for many of the fears, and one feels that until society gains control of its new technology these misgivings will continue. The computer has become the current 'whipping boy' of societies which are unable to keep pace with the rapid changes created by the introduction of new technologies; changes which can create a sense of alienation and anxiety.

One may conclude that it is not at all surprising that the computer has engendered anxieties, either rational or irrational. History has shown that many major innovations through the ages have created similar fears. The technological changes brought about by computerization have the potential for influencing the course of society far more than has any other mechanism in our history, and that the computer has been accused of causing physical, emotional and psychological change should not astonish us. However, the modern technologies of television and computers have perhaps created the only truly new fear, the one which accuses them of appearing to create an 'addiction' to their use, to the detriment of the psychological and social lives of the individuals concerned. Although 'television dependency' seems to be accepted as part of modern life, computer dependency still causes concern.

The influence of computer dependency upon computer systems

One of the major anxieties expressed with reference to computer dependency has concerned the effects of employing computer dependent people within the industry of new technology. Both Weizenbaum (1976) and Pearce (1983) presumed that the influence of such individuals could have negative consequences upon the systems they help to create. Not only was it assumed that they wasted much time by merely playing with the system, writing lengthy codes which only they could maintain, but also that their software would be unworkable or difficult to use, and their systems 'unfriendly'.

There is little doubt that many computer systems, whether commercial or domestic, do create significant problems for naive users, and if software is written and hardware designed with little thought for the abilities and limitations of the end-user then difficulties almost inevitably occur. Within the computer industry competition is usually so intense that user-trials are rarely carried out and software is not fully debugged before systems are sold. If such systems are initially designed by computer dependent persons, who have little difficulty understanding the nuances of programming and the foibles of computers, many will no doubt prove difficult to use by others.

Such criticisms may therefore be valid, but not all the blame can be laid at the door of the computer dependent employee. This research suggests that they form a very small proportion of those working within the computer industry, so the scale of their influence must be relatively small. Nevertheless, one may speculate that commercially available systems are designed almost exclusively by people who share many characteristics with the Dependents. Research has shown that the majority of programmers are field-independent and object-centred and find no difficulty working with new technology (see Poulson, 1983). If this is so, most of them, be they computer dependent or not, would probably find it difficult to appreciate the needs of the non-object-centred user who is fearful of technology; a mismatch in cognition and communication patterns is forged between the designer and the end-user. This does not excuse poor software design, but moves much of the blame from the computer dependent person to a far wider population within the computer industry. Further research would of course be necessary in order to establish whether differing cognitive biases could be one of the main causes of these discrepancies in communication between designers and end-users of computer systems, as suggested by this study.

Even if one assumes that the influence of the computer dependent person within the workplace had been severe, I believe that this is now less apparent than when described by earlier authors. Weizenbaum (1976), Thimbleby (1979), Pearce (1983), Frude (1983) and Levy (1984) all discussed the effects of those who worked with computers within large computer centres. Although a large proportion of the Dependents did use computers at work, some within universities and large computer industries, only one of the 45 interviewees within this group claimed to be 'hooked' upon the computing he undertook at work and owned no home computer. All of the remainder were dependent on the computing that they carried out at home on their own microcomputers.

This difference can be accounted for partly by the fact that the earlier authors tended to concentrate upon university populations, but also shows the passage of time. When these authors were making their observations home computers were rare and very expensive, therefore the easiest access to computer power lay within computer centres. The nature of computer dependency can therefore be seen to have changed over the years, as was confirmed by the interviewees. Although many of the Dependents had been 'terminal junkies' in the past, even back in the days of analogue computers, with the advent of the microcomputer and its peripheral equipment the computing undertaken at home had become of primary interest; the new computer centre was now the spare bedroom.

The Dependents' reasons for transferring their affections from a main frame to a microcomputer lay in the fact that they were now in total control of their computers; they did not have to queue on time-sharing systems, nor wait overnight to receive the printout of their labours; they could play with the hardware, the software and the memory, and hack into other systems without needing to be overly cautious about what they did. The power of the computer lay solely in their own hands, and those who still computed at work, on whatever type of machine, all preferred to confine their exploratory computing to their microcomputer. At home the user had total control over what was learnt, the speed with which it was learnt and the methods used; a freedom rarely found in the workplace. Most stated that their computing at work had become more efficient; they no longer had to spend excessive hours making up for the time spent playing with the system, but wanted to get home at night in order to carry out their own projects.

Extrapolating from the results of this research to the whole population of computer dependent people, one could assume that the trend to use home computers for the accomplishment of their preferred activities would be prevalent. Any detrimental effects caused by computer dependent persons at work can therefore be seen to have been greatly reduced in recent years by the advent of the microcomputer within the home.

The influence of computer dependency upon children

Although many authors have concentrated upon the indirect effects arising from computer dependency, others have suggested that the individuals could be directly and adversely affected by their intensive use of computers, and much of this work has concerned the influence upon the development of young children.

The consequences for children were difficult to assess as so few had taken part in the research, but some literature implied that the effects of computer dependency upon the young could be profoundly damaging. Waddilove (1984) suggested that children could learn nothing about the world from the isolated use of the computer; Levy (1984) believed that their use could be causally linked with introversion; Papert (1980) felt that programming in BASIC could lead to a restriction in language and expression; and Weizenbaum (1984) suggested that games players could become psychically 'numbed'. Brod (1984) elaborated further, suggesting that children's ability to learn would become 'distorted', and that they might develop an 'intolerance for human interaction' by comparing people unfavourably with the fast, responsive computer. He saw the use of the computer becoming a refuge from stress, preventing the development of a well-rounded personality by cutting the child off from other activities, and leading young games-players to become non-cooperative and unimaginative. Such beliefs conflicted with those cited by other authors (see Chapter 10) suggesting that there is much disagreement in this area.

The young people who were interviewed did devote most of their time to computing at the expense of other activities and, like the adults, sometimes did use it as a release from stress and as an escape from social interaction. However, such behaviours did not appear to have been causally induced by the computer. As with the adults who took part, these intelligent youngsters also tended to have been very shy from early childhood, and had always been absorbed in activities, often of a technological nature, which did not require the participation of others. They tended not to be interested either in physical or social pursuits, but preferred to indulge in activities which were intellectually challenging and furthered their knowledge of the technical artefacts of the world. Most agreed that their studies might have suffered, but they felt, perhaps incorrectly, that their expertise with computers would compensate adequately for anything they might have lost. The computer had not altered their personalities, merely focused their interests.

One may assume that computing had cut them off somewhat from making more effort to be sociable, but this is not what they believed. All claimed they would still devote the majority of their time to solitary pursuits, and their past histories of hobbies seemed to bear this out. This was also confirmed by the teachers interviewed for the Schools' Survey (Chapter 5) who believed that the youngsters whom they labelled as computer dependent had social problems before using computers and had been very positively interested in new technology. Neither the personalities of these children nor the interviewed Dependents appeared to have changed; they had merely found an outlet for their talents and a refuge from the situations in life which had previously caused them stress.

One may draw conclusions from this study similar to those of Simpson (1983), who recognized that 'computer addiction' was a response to deficiences elsewhere in life and deduced that such an activity could be in the child's best interest. Similarly, Nicholson (1984) believed it to be symptomatic of needs not met elsewhere, and in addition felt it to be preferable to many other activities and no more socially isolating than watching television. If fostered, one could assume that if the young computer experts were encouraged to share their talents with others at school, that not only would their confidence and self-esteem increase but so too would their social skills; factors rarely addressed directly within an educational setting.

On the basis of this present study one may conclude that people should have few fears that use of computers damages the personalities of the young; computer dependency only appeared to have occurred for those children with a predisposition for technological activities and with existing social difficulties, both manifested very early in life. Aspects of the Dependents' childhoods were shown to have been instrumental in the development of their object-centred biases, and had been laid down long before they arrived at school or used a computer. Parents and teachers should not assume that any child who uses computers will become fixated upon them to the exclusion of other activities; the vast majority will not, and those few who do will perhaps gain more benefit from this activity than from any other. Perhaps we should be more concerned about the multitude of children, especially girls, who at best are disinterested in new technology, and at worst remain fearful or hostile.

Computers as dehumanizing agents

Although a few authors admitted that computers could act as liberating aids to thinking and the development of logic (see Eisele, 1981; Pateman, 1981; and Turkle, 1984) others revealed fears about the way they felt computers would be viewed by those less knowledgeable than themselves. Because the computer is able to mimic human intelligence to some extent, a few authors felt that any anthropomorphization of the computer could result in it acting as a dehumanizing agent. Such an effect was thought to occur because of the current trend for designing 'user-friendly' software, which could lead some to believe that they were interacting with an intelligent 'person' rather than a mere machine.

Boden (1981) warned strongly against designing quasi-human, user-friendly systems, believing them to encourage the 'alienating substitution of computerized companionship for human conviviality'; thoughts also echoed by Plum (1977) and Fitter (1978). Thimbleby (1979) believed the projection onto an object gave it 'symbolic power', and Frude (1983) and Simons (1985) felt that people would be all too willing to accept computers as 'human', with a resultant lessening in need for real human interaction.

Weinberg (1971) felt that our manner of thinking would need to change in order to fit that of the computer, while others believed that social mechanisms would alter. Sprandel (1982) saw communication patterns being reduced only to those which were obviously and directly expressable, thereby removing the subtle nuances of the human interaction, and reducing the user to living in a self-created, computer-centred world. Meek (1972) felt that people would credit the computer with the human attributes of fairness, trustworthiness, reliability and veracity, such as Weizenbaum (1976) found when people interacted with his *Eliza* program. Designed to parody the role of a Rogerian therapist, users found *Eliza* to be a medium with which it was very easy to communicate, to such an extent that the program was seriously considered by psychiatrists as a valid medium to use within treatment. Weizenbaum found such responses 'corrupting', believing that users were being reduced to the level of machines.

People of many levels of expertise do talk to/at computers when they use them. Scheibe and Erwin (1979), from experiments looking at the verbal comments from humans when interacting with computers, concluded that their results showed 'evidence for the temporary personification of the computer', to the extent that some subjects 'asked if a real person was not in fact making the responses for the computer'. The researchers suggested that the computer would come to be viewed 'not only as person-like but god-like' in the eyes of some, to the point of the 'emergence of a new personality type which is the human reciprocal of the computer'. For similar reasons, Schwartz (1983) felt that transference feelings could be evoked by the machine, and Boden (1981) stressed that all systems should deliberately retain an element of 'threat', and that any program which made the user 'goggle-eyed with astonishment, should be strictly curbed'.

The authors of these theories were in the main voicing anxiety about the effects of computers in general rather than upon computer dependent users in particular. However, one could assume that they would view the Dependents as exhibiting the very characteristics they fear will occur to other users of computers, with dependency upon computers the extreme example of what may happen if their theories were realized. Their suppositions can only be replied to in the light of the Dependents' beliefs and behaviours, and they were found not to apply except at the most simple of observational levels.

It was true that many of the Dependents did talk to their computers and viewed them with affection; sometimes naming them or describing them as friends or colleagues. People often name their boats and houses, and there is surely little difference between a person abusing a recalcitrant car, or praising and pampering a beautiful and reliable one, and those who talk to their computer. Perhaps any object which is seen to 'respond' will be spoken to. People frequently thank automatic doors and speak lovingly to their plants; pet-owners talk endlessly to their faithful, non-human animals, who are often felt to understand every nuance of speech. The computer actively responds both visually and aurally to the user's input, and depending upon one's skill and familiarity may be viewed as threatening or cooperative and spoken to accordingly. Merely speaking to an inanimate object is very common but does not necessarily suggest that the interaction is dehumanizing or abnormal. It may be little more than the remnants of the child's need to monitor and guide their actions through commentary.

This research demonstrated that the computer did act as a rewarding medium with which the Dependents worked and played, and for many a type of 'relationship' often did develop between the human and the machine. A computer, like a car, a boat or any other piece of intricate equipment, has its quirks and vagaries which have to be understood and responded to sympathetically; therefore it is not unusual that a kind of rapport may develop as the user becomes familiar with it. However, because someone speaks to an inanimate object or a non-human animal and appears to have a relationship with it does not imply that it is viewed as human or even quasi-human, even if it appears to respond in such a manner.

That computer professionals and artificial intelligence experts have become concerned that the human-computer interaction might appear quasi-human seemed somewhat bizarre. If computer software gives the impression of responding in a human-like manner, it is for the very reason that a *human has written the program*. It is therefore of little wonder that a computer can appear to mimic a human/human interaction; that is just what it is. The software designer is communicating with the end-user. The computer did not design the interaction, and even when computers are used to program other computers the original software will have to have been designed by humans. Why Weizenbaum was surprised by users' reactions to *Eliza* may perhaps be seen as more astounding than their reactions to the program. He had deliberately attempted to mimic the therapist and had admirably achieved his aim. (His reaction may say more about his attitude to Rogerian techniques than to the computer users' responses, however.) Whether or not machines should be used for such tasks is a separate issue, but Weizenbaum had created something which he felt was being sorely misused; a veritable Frankenstein.

As new technology advances and microcomputers become more powerful there is no reason at all why programs should not communicate in terms which are simple to understand; there exists now the capability to remove the 'computer-speak' from the users' interaction, replacing it with a friendly jargonless interaction using natural language. However, even software which is gratuitously obsequious and familiar will not, I believe, encourage users to view the computer as human, just as a computer which says 'Hello Bobby' to a little boy will be thought any more human than would ET; it so patently is not. This applies especially to the Dependents who have an extensive understanding of the basic architecture and techniques of programming and are well aware how software is designed. Turkle (1984) found that even young children, who believed that computers were 'alive' and intelligent, were well able to distinguish between humans and computers and tended to describe people as essentially what computers were not; although not necessarily in the same manner as did the Dependents.

In spite of any familiarity between the Dependents and their computers, none of them seriously viewed the computer as human-like; in most cases they were seen as antitheses of each other and computers were appealing for just that reason. Where computers were viewed as logical, predictable and non-judgemental, the human was considered to be illogical, erratic and critical (see Chapter 8). It was the very difference between computers and humans which made them attractive to the Dependents. If the authors fear that a human-computer interaction will be preferred to a human-human one, their fears are justified to some extent; however, this is surely not the fault of the computer but of society and its members. It was not without some justification that the Dependents turned to the inanimate world for satisfaction, but I believe that few will follow their example.

The interviews which were carried out and the comparisons made between the Dependents and the control groups indicated that computer dependency and the development of a relationship with the computer did not develop in a vacuum. If they may be described as prerequisites for the development of computer dependency, the results which differentiated the Dependents from others indicated that the individuals invariably had to have experienced unsatisfactory relationships with others throughout their lives. These appeared to have been caused by parenting methods which were undemonstrative and distant, involving love conditional upon the performance of deeds. In addition, the person had to be highly intelligent, with an object-centred, field independent cognitive style, and have an inquisitive, even insatiable, curiosity about science and technology, as reflected by their early choice of hobbies and activities.

Few will fulfil such requirements. The majority of the population will either remain disaffected or uninterested in new technology, or will only view the computer as a means to an end; a means of carrying out specific tasks, whether for work or play. Without the desire to explore the system and the intricacies of the working of new technology, computer dependency will be very unlikely to develop.

The design of user-friendly systems has been thought to lead to the personification and humanization of computers. However, if the computer dependent person is thought to exhibit the most intense level of such interactions with computers, this research has demonstrated that this belief is fallacious. It could even be assumed that, if computers do become extremely user-friendly and easy to use, then computer dependency may no longer exist at all. 'The *Macintosh* is just silly, I hated it', a comment from one of the female Dependents echoing the views of the majority. The *Macintosh*, perhaps one of the most advanced and friendly microcomputers, was viewed by many Dependents to be merely an uninteresting 'black box'. They did not feel that they could explore it or learn from it; it was merely a tool. To them, such a machine removed all of the intrinsic interest and excitement of discovering the capabilities of the technology the hard way, in a way in which they could really begin to understand and master the system. Similarly, the modern locomotive and car are perhaps of far less interest to real enthusiasts than their older counterparts. Although interest still exists for the modern versions, it is perhaps more for their extrinsic rather than for their intrinsic uses and qualities.

Many of the Dependents learned machine-code programming and studied the computer hardware in order to increase their understanding of the technology, neither of which are essential if one wishes only to achieve results. Many successful users of computers probably have no notion of how a computer actually works, nor do they wish to know, any more than many car drivers have a desire to understand more than where to put the petrol, oil and water. Such knowledge is not essential for the successful use of a machine if it is easy to use; when serious problems develop an 'expert' can be consulted. The Dependents did not follow this philosophy, however. They enjoyed the problems, 'bugs' and the jargon which helped them feel part of a knowlegeable elite, and were not deterred by the poor manuals which created the need to learn through exploration. User-friendly systems which are able to assist the user without recourse to learning alien terms and languages could be seen therefore to spell the death sentence for computer dependency. The Dependents believed they could not feel affection for a black box about which they could learn little; it was merely a tool, not a toy. Perhaps if user-friendly computers are used within schools, for activities which are extrinsically useful rather than just for the study of computer science, such relationships will rarely exist and the disaffected will come to view computers as useful, non-threatening tools.

That the Dependents had 'relationships' with their computers and preferred that interaction to that with other humans cannot be disputed, but rather than viewing the computer as another human it became apparent that the Dependents' interaction with the computer was more narcissistic than anything else. When programming, they were interacting with the ideal partner who understood them fully; they were communicating with themselves. The computer became an extension of their own personalities, exhibiting the same logical structure that they too applied to the world.

Although some may view this as an unhealthy way of conducting one's life, all of us live a narcissistic life as far as we can. We choose newspapers which reflect our opinions, watch television prorammes which appeal to our tastes, worship with those of similar beliefs, and are probably even more discriminatory in our choice of friends. By choice, most people would prefer to surround themselves with like-minds, and the types and levels of discrimination present in the world indicate that tolerance of other groups is not commonplace. Narcissistic activities abound, and the Dependents' preference for an interaction with the computer was often no more than this. Instead of continually facing the frustration of interacting with humans who did not share their modes of thinking and interests, they created a world which did and which fulfilled their needs. Even when 'interacting' with the writers of commercial software they were communicating with those of like-mind, just as they were when networking.

The Dependents saw little similarity between their interaction with the computer and with most humans. The computer did not interrupt, did not misinterpret or get upset, was unambiguous and straightforward, and for these reasons was easy to deal with. Human communications do lead to misunderstandings, especially for those who are inexperienced in sharing themselves fully with others. One of the Dependents even experienced difficulties when watching television plays, as the nuances and the interrelationships were incomprehensible to him. Kidder (1981) illustrated such problems with the description of a wife, who when she asked her hacker husband if he would like to carry in the groceries, became upset when he said no; he assumed the question allowed for a choice of replies. Hollands (1985) recognized these difficulties within interactions, and stressed that partners needed to be aware of the interpretations which can be placed upon the spoken word by very literal and logical scientists.

Although it has been feared that the computer would alter modes of thinking and communication patterns, the interviews with the Dependents and some of their partners showed that this had not occurred. Subtle nuances of speech were not part of the vocabulary of the Dependents, nor were they as fluent or as able to express themselves as easily as their partners, with whom communication was often fraught with misunderstandings. Their communication with computer software, either written by themselves or by others with similar cognitive styles, was in the language which they understood; the Dependents only had to alter their style of thought and communication when interacting with humans who differed from themselves.

From these results, one may therefore assume that as far as the Dependents were concerned, little had changed except the opportunity to interact with an 'intellect' similar to their own. Although some of the wives found their Dependent partners' activities to be incomprehensible and alien, even viewing the computer as a rival, they need not have viewed it as the equivalent of another woman; at most it acted as a reflection of the Dependents' own personalities. Although many of the Dependents did view the computer as a colleague and a friend, they did not see it as human but rather as 'superhuman'. The fear that computer use could lead to dehumanization and cause people to view others as machines (Papert, 1980) seemed unfounded. It was because humans had shown themselves to be unlike machines and unpredictable which had created the Dependents' interest in artefacts in the first place. Their enthusiastic pursuit of their interest did not change their basic characteristics, personalities or orientations but merely allowed them to flourish.

Computers as desocializing agents

The most serious effects which were found to be caused by the Dependents' use of computers were within close relationships; not an aspect stressed by many authors writing about computer dependency. Some unmarried couples had separated, and a few marriages were showing signs of strain although they still appeared stable. Problems did not occur within all of the marriages however, only within those where the partner expected to share interests and large proportions of time with their spouses. Such difficulties had been recognized by a few others, although their fears did not reflect what was truly occurring.

> 'Already spouses are having to compete with computer systems for attention and affection, a situation which in the past has occurred in connection with other types of artefacts but which now has acquired fresh, sinister overtones.'
>
> Simons (1985)

The use of the word 'sinister' by Simons reflected the view of other authors who suggested that the computer had a more powerful influe than objects hitherto, to the point where they could act as the pr love and affection and as substitutes for personal interacti

> 'It may be possible for intelligent machines of the future to su intellectual stimulation or instruction, but also domestic ar

social conversation, entertainment, companionship, and even physical gratification.'

Firschein *et al.* (1973)

Although perhaps stated with some flippancy by Firschein *et al.*, both Frude (1983) and Simons (1985) seemed to have taken the last point with some seriousness, even envisaging the day when a human-like, robotic doll would act as the perfect sexual substitute thereby removing the last vestiges of need for human contact. Similarly, Levy (1984) in his observation of hackers, believed that computing 'had replaced sex in their lives' as women were seen to be imperfect and unpredictable, and Laurie (1980) similarly suggested that a machine which appeared to be alive, could communicate and was controllable, could also have the same effect: 'perhaps computers will replace sex altogether'.

There was little doubt that the sexual relationships of some of the Dependents had been adversely affected by the advent of the computer, to the distress of both partners. Couples no longer went to bed at the same time, and wives withdrew their favours when not prepared to be woken in the early hours of the morning. The impression was not gained, however, that the Dependents no longer loved their partners. Sex was important to both but the manner in which it was carried out, fitted in between two bouts at the keyboard, was intolerable to a few wives.

Computers will indeed threaten humanity if they become universal substitutes for sex, but this seems unlikely to happen, even to people like the Dependents who experience social difficulties. The majority of Dependents were in favour of marriage, and earlier results showed that the Dependents and the two control groups yielded similar proportions within each of the marital categories (see Chapter 4). Sexual relationships and marriage are aspired to by the majority of the population, and the conversations with the Dependents revealed that they were as willing and able to fall in love as the next person, that they enjoyed their relationships, and desired the security offered by marriage. Even if some of the men had to make signficant efforts in order to win their partners and overcome their fears of women, as had been suggested in Chapter 10, many were successful and usually considered themselves to be happily married.

Even those within marriages which were dogged by problems caused by their computing did not wish to end them, and only the divorced female Dependent appeared to have removed herself from the arena. She appeared disaffected with men in general and was surviving quite happily without them. Even if more women do become dependent upon computers this is unlikely to affect the progress of the human race, as women have usually been married before they draw such conclusions.

One might suggest that the young, single Dependents will be unlikely to marry as they will not make time to mix socially and meet members of the opposite sex. Once again the evidence suggested this to be false.

The adult Dependents did not alter their personalities with the advent of the computer; most had been extremely shy since childhood but the urge to find a partner had been strong, and the young Dependents intimated that they too were prepared to take time away from their computing in order to find a mate when the time seemed appropriate. One could suggest that it is not only Dependents who make such rational decisions about their future relationships and compartmentalize their lives in this manner.

One could perhaps foresee that the only real threat to marriage and the procreation of the species, indirectly caused by this type of computing, will occur if women become more adept at judging the personalities of their men before marriage. However I feel that even this is unlikely to occur. Although the women did have an inkling of 'what they were marrying', as suggested by two husbands, the personalities of their men were found to be extremely attractive before marriage, perhaps because they were such individualistic characters. Only after marriage did they both realize the extent of their differences, although only the women appeared to suffer. Their needs demanded the sharing and caring presence of their men; needs which the men neither appreciated nor seemed able to fulfil. The men were tolerant of any difficulties within the relationships, mainly because they were in complete control of the situation and were meeting their own needs. Counselling would no doubt have been very beneficial to enable each to understand the other's point of view, whereby both could adapt their behaviours, but none seemed to have taken such a step.

Many of the Dependents failed to understand their wives' needs for their company, or to understand that they were expected to add more to the relationship than merely provide a salary, a home, and children. The men truly felt themselves to be happily married; they were content, they demanded little, their expectations from relationships were limited, they felt they fulfilled their side of the bargain and had little concept of any loss. For most, their marriages had provided the best relationship in their lives and they felt complete. Most had no desire to end the relationship, let alone to start another, and were quite horrified by their wives' threats and jealousy shown towards the computer. As the Dependents' computing was merely an extension of their own thoughts, their wives' reactions seemed incomprehensible to them. Many had deliberately attempted to interest their wives in new technology, but most attempts failed; their wives, even if they had made serious efforts to study programming as two had, could not appreciate that a technological artefact could be viewed as any more than a tool of limited application.

Although differences in interest levels between the sexes had not been explored and no papers were found to this effect, it seems from observation that it is more likely to be males who tend to have one serious hobby while females are more eclectic in their interests. It is not unusual to note that there are 'groupies' attached to many male activities, females who support the activity but are rarely a part of it, while the opposite is rarely found. Women have perhaps traditionally accompanied their males in order to spend time with them, and no doubt with the Dependents' previous activities there was more opportunity to share a little of the pleasure engendered even if this only meant chatting in the garage. Computing demands one's whole attention and is not an activity conducive of simultaneous conversation. If the wives had no wish to share the activity directly their presence was a hindrance, and when engrossed at the keyboard the Dependents were said by their wives to be scarcely aware of the presence of another.

The lack of communication between couples was profound in a few cases, with each partner appearing to view the other with incredulity; neither could understand the other's point of view, nor could they appreciate why each other's needs were so antagonistic to their own. The husband failed to understand why all were not excited by new technology, the wife that someone could live with almost no meaningful personal interaction. However, the wives could not state that their Dependent partners' personalities had changed. Their descriptions showed that their husbands had never been good communicators, had always found it very difficult to express their emotions and feelings, had almost invariably been heavily involved with other hobbies in the past, and rarely participated in family life to the extent that their wives desired.

This research had shown that the personalities of the Dependents appeared to have been well formed early in life; the computer had merely given them the medium by which they could express themselves fully. If the Dependents' personalities had not changed, two important factors had: their behaviour had altered in as much as they now devoted more time to computing than to their other activities, and the attitudes of their wives had altered. The tolerance shown towards their husbands' previous hobbies could in some situations no longer be sustained. Where a husband had only occasionally spent all night working on some activity, the advent of the computer with its availability, accessibility and catholic applications had led to these occurrences becoming frequent. It was not difficult to understand the wives' distress. Although having lived with their husbands' previously hobbies with a certain degree of tolerance and restraint, in some cases the overwhelming interest in computing had gone beyond the realms of what could be considered acceptable.

Although within marriage both had obtained the security and emotional support they needed, the Dependents were more easily satisfied; their needs were fewer and more straightforward and they had realized their aims. Their wives, one could assume, had expected to enhance and alter their partners' social reticence and inhibition, and even if this had occurred to some extent, the advent of the computer into the home had removed any ground they had made. The Dependents' basic personalities had once more flourished and become entrenched, and the wives felt deserted.

Neither partner believed the other was doing anything useful; both sitting together watching television and computing were seen by the other to be pointless and time-wasting activities. But only the wives were miserable, as their preferred activities demanded the participation or the company of other individuals. Their main hobby could be seen to be 'people', with their husbands understandably being the chief object of that interest; an interest which often proved incomprehensible to their partners. Although both had married for security and companionship these were interpreted differently. Stalemate had been reached and the problems seemd insoluble to them, although most seemed resigned to the situation.

The results from this research, although demonstrating that problems did occur within social relationships, highlighted the fact that problems did not exist where none had occurred before. The Dependents had shown histories of poor social relationships, even within their marriages, long before the computer had entered their lives. Their use of computers caused additional difficulty simply because more time was devoted to this activity than to any of their previous hobbies. Although their social inhibition had often caused them distress early in life, the interviewees demonstrated that they had a firm grasp upon their limitations and personalities and had adapted themselves to lifestyles which removed them from most of the stresses caused within relationships. Comparisons made between the groups' Self and Ideal Self scales (see Chapter 8) demonstrated that, although all three groups would have wished for greater social skills with the Dependents more removed from their Ideals than the controls, there were no differences between the groups for their general contentment with life and with themselves. The Dependents had established very satisfactory lives which differed greatly from the suggestions made by many of the authors of the literature.

The authors who expressed concern about the social effects appeared to imply that the use of the computer was actually altering personalities and changing the extrovert into a non-communicating introvert; beliefs which were shown to be unfounded by these results. What was of interest was why the authors had overstated the case, and why they had chosen to write about the damaging effects of computers in such a manner.

Observations upon the reactions of others to computer dependency

Although some authors had not written specifically about computer dependency but about computer use in general, many implied that those who used computers for anything other than for purposeful problem-solving could be adversely affected by the interaction, especially if this developed into some sort of 'relationship' between the person and machine. Some of the authors were highly-respected academics, their arguments appeared well-founded, and they tended to imply that people needed to be cured from their attachment to an anthropomorphized machine. It was therefore expected that the effects they mentioned would be found to exist for those who took part in this research when the syndrome was studied in more depth. Their beliefs had inspired this research, and it was of some surprise to discover that their deductions were only found to occur on the most superficial of levels.

Pleasure obtained from an impersonal other had seemed to promote the authors' concern. However, as many media have similar power, one might ask whether the same authors were also concerned with the effects upon people who positively identify with fictional characters who appear on the cinema and television screens, upon the stage or within books. Such media are designed specifically to create artificial realities and are often employed to lift the recipient out of the mundaneness of everyday life. Is a computer language so different from that created by the pen of another type of author? Is there something special because the use of the computer involves a two-way interaction which makes it more threatening than others? Should people be protected from the output from a computer screen by those who would perhaps fight for freedom from such censorship within the press and the cinema?

It was of interest that the very authors who suggested that the masses should be protected from friendly computer systems were often those who themselves had spent many years working with new technology. Had they remained unscathed, or were they concerned that the negative effects their computer-interactions had had upon their own personalities would affect others similarly? Or were they similar to those who spent days watching pornographic films in order to censor them and protect those less morally strong than themselves? These questions cannot be fully addressed here, but the results from this research indicated that their fears were groundless. The Dependents were no more changed by their interaction with computers than most of the authors would probably admit they had been.

The authors cited were not alone in assuming that computer dependency was damaging. During the course of this research various press reports stated that a psychiatrist working in Scotland was actually treating people who had experienced this syndrome. A long telephone conversation with this psychiatrist revealed that during the course of four years he had treated eight males, referred to him by general practitioners, who were 'showing signs of psychological problems and neurotic tendencies who were not responding to tranquillizers'. Through history-taking he discovered that they all used computers 'excessively', from 12 to 14 hours per day, and he had 'established that computing was the main precipitating factor of their illness'. All were 'considered to have been perfectly normal, with one slightly introverted but not eccentric' and all were 'exceptionally intelligent'. Five were youngsters who had been taken to their doctors by their parents because of psychosomatic illnesses, lack of appetite, lack of interest in 'normal activities', and because they lacked 'cheerfulness'; the three adults appeared to conform to the descriptions of the Workers in my research and were using computers for work-related activities for extremely long hours. None of these individuals was concerned by the length of time they spent computing, and the adults were said to have put up a great deal of 'resistance to the idea that their computing could be bad for them'.

As computing was believed to be the cause of their problems, rather than a symptom or an expression of their personalities, the psychiatrist saw his aim to 'modify their behaviour' by reducing the hours they spent on the computer and by encouraging them to take up other hobbies. This he did using 'psychotherapy and hypnosis to induce deep relaxation' and he seemed to be successfully reducing the computing hours of his patients. Nevertheless the treatment was time-consuming, often taking over 18 months and two people had been treated for over two years. The psychiatrist's reason to explain why these patients had become so interested in new technology was that the behaviour was 'reinforcing', 'very satisfying and pleasurable'. If computing was the realization of needs not met elsewhere, as suggested by this research, one hopes that other problems were not caused by weaning them away from this activity if nothing was offered to replace it.

The psychiatrist's beliefs perhaps differed little from those of earlier authors writing about computer dependency. A behaviour was observed which appeared abnormal and the computer was blamed for causing the behaviour; the cure was therefore to prevent the behaviour continuing or or from occurring in the first place. This could be achieved either by cutting down the hours spent at the keyboard or by designing systems which did not lend themselves to being treated anthropomorphically.

One can image how the conclusion could be drawn that computers were to blame for the creation of this new phenomenon. Observation of 'terminal junkies' within computer centres would have shown people, mainly male, who seemed totally unconcerned with the normal events of life, such as eating, sleeping and dressing conventionally, who were perhaps unusual in that they had no desire for breaks from their labours, who were not intimidated by new technology and actually enjoyed working with computers, even to the extent of preferring them to people. Such behaviour is 'abnormal', especially in a society where so few seem to enjoy their work, where many do not like interacting with new technology, and where most would far rather talk to another human being than attempt to come to grips with a computer.

Observation of the activities of the computer dependent person would give the distinct impression that the features of computing were the 'cause' of the syndrome. Use of the computer offered the Dependents reinforcement, excitement, intellectual stimulation, the feeling of power and potency, unlimited challenges, prestige and self-confidence, to name but a few of the reported positive effects as discussed in Chapter 10. But mere observation inhibited a greater understanding of why the Dependents had developed this passion for computing. It had not arisen in isolation from previous needs, but was the culmination of those needs, as prior knowledge would have shown.

No doubt the computer dependent person would have been unlikely to draw the attention of observers before computing had become of interest. Their previous hobbies and preferred activities were invariably solitary, and when at work they probably were less likely to interact with their colleagues than others. Only when they started to use terminals in computer centres, under the scrutiny of others, would their interests have been open to view. Colleagues, who previously may have appeared withdrawn, quiet, hard-working and sensible, now seemed to have developed irrational behaviours. They began to work longer hours, spent much of their day playing with the system instead of producing usable output, missed lunch breaks, and appeared to be more concerned with computing than with their families or the things which occupy the interests of the majority of the population. They appeared to have altered significantly, and were now viewed as 'weird' and 'eccentric', and labelled as 'computer junkies' or 'compulsive programmers'.

The computer dependent individual's opinions were not sought by most authors, and because of the social difficulties invariably experienced by such people they would probably have had great difficulty in describing their fascination with computers, themselves or their desires in an unstructured, conversational manner even if they had wished to do so. The authors therefore had to rely upon their own interpretations of their observations.

I was fortunate to be given the opportunity to do more than merely observe the Dependents' interaction with computers and was able not only to make comparisons between them and their peers but also to interview them and their partners. It was only after this that a different picture emerged.

We should not be surprised that computer dependency occurs

Although causality cannot of course be firmly established, the results which differentiated the Dependents from the control groups and the information gained during the interviews suggested that there were more features within the person which were significant to the development of computer dependency than within the computer itself. The computer was central, but if it had not existed (and when it did not exist) in the lives of the Dependents the majority would have been engrossed in activities of a very similar nature which offered the same types of intellectual rewards (see Chapter 7). At an early age the Dependents had developed object-centred interests and had dispensed with much human interaction.

Experiences throughout their lives had shown them that interaction with humans was painful; a lesson learnt from humans, not computers. The early experiences of the Dependents gave them their initial instruction in human behaviour. Most discovered that they would be rewarded as people by the performance of actions and deeds, not by merely being themselves. Their parents were cool and aloof, but encouraging, in the manner of the middle classes, and many Dependents did not feel that they had been loved unconditionally. They did not learn the art of loving communciation as they had no adequate role models, and humans came to be seen as untrustworthy and unreliable. Their initial needs for human love and acceptance had been strong and normal, and many, even as adults, still felt extremely hurt by the lack of outward affection in their childhoods (see Chapter 8).

Humans were unpredictable, but the Dependents learnt that the inanimate were not. Logical outcomes tended to be the result of the manipulation of objects; achievement and self-fulfilment were forthcoming, and the Dependents seemed to have learnt to be able to cope with their situation by becoming reclusive achievers from a very young age. Through the manipulation of artefacts and the internalization of ideas they grew as humans into successful members of society. They were intelligent, and had developed cognitive biases and abilities in the fields of mathematics, science and technology which were valued by others. Nevertheless, they seemed not to have developed their full emotional potential, and most never achieved any measure of true understanding of people of either sex. Although their dependency upon people still existed, for many this need remained hidden and unavailable to them, even to their partners whom they loved for loving them unconditionally. They did not know how to reciprocate fully when affection was offered, and one could see the 'sins of the father' being passed on through the generations as they themselves became non-communicative partners and parents, however hard they attempted to alter this.

Artefacts of one type or another had invariably been at the centre of the Dependents' interests, allowing them to gain power and control over sections of their environment. This, coupled with their interests in philosophy, science and science fiction, allowed them not only to follow the doctrine of intellectualism, but also to remove themselves from the present and the mundane. The advent of the personal computer fulfilled their need for knowledge and the pursuit of logic and pure reason. It was the first piece of technology which had allowed them to interact on a seemingly intelligent, conversational level with an equal other. This research has shown that those who desire this type of objective interaction have found solace and fulfilment when working with new technology, and that invariably they are the ones for whom social fulfilment had been lacking.

The Dependents' lives had in many instances been traumatic and such traumas had invariably been caused by humans. Their parents, perhaps ultimately responsible for their children's later abilities to share and love, tended to withhold their affection. Through their familial conditioning the Dependents had been trained to perform tasks in order to gain affection, and even in adulthood when offered unconditional love by their partners could not stop performing and learn to trust others implicitly. They gained satisfaction from the performance of deeds; it had become part of their personalities. The manipulation of objects and ideas was central to their lives, and a relationship, however secure, could not fulfil their basic needs. To quote Catt (1971):

> 'The computer is today's philosopher's stone, the supreme object of alchemy. But whereas the philosopher's stone offered only wealth, the computer seems to offer spiritual aid as well.'

Computing had offered spiritual aid to the Dependents and had brought them great comfort and stability. They were functioning normally and often very successfully within society and were not asking for cures from their activities. They were content and fulfilled, and had achieved what could be considered for them to be the 'ultimate interaction'.

It is not difficult to see why some people opt out of social interaction and prefer the pursuit of a spiritual or intellectual life. Humans create more difficulties in the world than any inanimate object, sub-human species or natural disaster. Humans cause anger, misery, famine, war, divorces, murder, genocide, etc.; machines may only be involved indirectly, being controlled by humans. However, a life without friendship and human interaction is considered undesirable, and a 'relationship' with a machine somehow unhealthy. Many continue to strive to prevent the former occurring, and persevere in spite of any misery and suffering which ensues, often by escaping the pain caused by relationships by turning to another relationship. Humans are used as a buffer against other humans. However, some do opt out of this cycle. They may turn to religion, to alcohol, pets, gardening or fishing; still others to interactions with computers. All have a method of coping with the difficulties which arise in life, but the thought of escaping to a computer is anathema to many. It seems sinister, frightening, and smacks a little of the dehumanized aspects of Orwell's *1984*.

Suffering is part of the Judaeo-Christian tradition, and perhaps for this reason a form of escape which inhibits personal pain and frustration, but which appears totally self-indulgent, is viewed as unacceptable. For some people relationships are easy, computing difficult, for others the reverse is true; but one seems to be considered more acceptable than the other. Nevertheless, the results from this research suggested that the Dependents were no more unusual than many other people who prefer to interact with the inanimate than people, many of whom are respected for their activities and contributions to society.

Do computer dependents differ from other enthusiasts?

Other authors had considered the Dependents and their activities to be unusual if not subversive, although this was not the impression gained during the interviews. To me they seemed no more unusual than many of the dedicated academics with whom I am surrounded, not far removed from other hobby enthusiasts, many of whom are admired for their dedication to their work or interests. Much of the Dependents' time was spent undertaking activities which were not extrinsically useful and perhaps for that reason their activities were considered unusual; but is dancing useful, or singing or playing football, all of which can become means of earning a living for some? Are local and national sporting events not harbours for the encouragement of competitiveness, enthnocentrism and chauvinism; hardly as innocent as solitary computer programming but rarely criticized? Are train-spotters, stamp-collectors, bird-watchers, radio-hams or pigeon-fanciers any more unusual than the Dependents, all of whom are undertaking activities which must seem distinctly bizarre to the majority of the population and usually of no 'use' except as a means of recreation and play for the individual concerned?

The Dependents are not the only type of people to have little need for human interaction, or who dedicate themselves to other activities. Many who live the monastic life, or the life of the intellectual, artist or musician, often have little need for others and are frequently admired for their dedication. Dedication to any activity invariably entails the sacrifice of worldly needs in the pursuit of particular ends, and perhaps such activities are usually undertaken by those who initially have no interest in the more mundane aspects of life.

It could be hypothesized that all who seriously dedicate themselves to one activity, to the exclusion of others, have equivalent backgrounds to those of the Dependents. Perhaps those who have no need for hobbies or who are found to be eclectic in their interests differ psychologically, emotionally and socially from the dedicated enthusiast, and perhaps more detailed research comparing the personalities of the exclusive and the eclectic hobbyists would reveal these differences. There has perhaps always been a proportion of the population who have traditionally been machine-oriented, some of whom gain great pleasure, success and prestige from their relationships with the inanimate. This theory was confirmed to some extent by the observations of the racing driver John Surtees, whose sentiments and history appeared similar to those of the Dependents. Few would suggest that Surtees' relationship with cars had not only been of benefit to him but also to many thousands of people who have watched him race.

> 'From the time I was a schoolboy and spent every spare moment tinkering with pieces of wood, and later a motorcycle engine, I often found it easier to relate to a piece of machinery than to some people. The relationship that a driver or rider must have with his machine on the track is a very special one, sensitive at times, dominating at others. Just as some people have experienced how bricks and mortar, timber and tile in the shape of a house can have a real character, and virtually talk to you, so the same can apply to a piece of machinery . . . As far as I am concerned, my 507 will always have a very special place in my affections.'
>
> Surtees (1988)

The results from this research suggested that the Dependents did not differ greatly from enthusiasts interested in cars, motor cycles, amateur radio or model aeroplanes. Indeed the results from the Leisure Activities Inventory and the interviews suggested that these activities were also of great interest to the Dependents before the computer came into their lives. Letters received from the wives of enthusiasts interested in the above mentioned hobbies (see Chapter 10) suggested that their partners' personalities were similar to those of the Dependents and that the effects of their hobbies within the family were much the same as within those of the Dependents. Perhaps such enthusiasts, if showing the Dependents' level of interest in philosophy and science fiction (two of the main factors which differentiated the interests of the two computer-owning groups, see Chapter 7) will also graduate to the computer and share the same type of enjoyment and stimulation, which for the Dependents superceded those gained from their previous activities.

The aforementioned hobbyists showed similarity to the Dependents, no doubt because they were all machine-oriented in their interests. One could therefore conclude that the Dependents behaviours were not abnormal, nor even untypical. However, others also carry out activities which remove them from the social sphere and often depend upon isolated dedication; artists are such a group. Did they also share similar characteristics with the Dependents and other machine-oriented persons? Do all dedicated people in fact share similar traits? In order to examine whether this theory had any substance an additional small study was undertaken.

Interviews were conducted with two people who seemed to gain similar levels of pleasure from their task-centred activities as did the Dependents. Not wishing to compare like with like, people who practised an activity which on the surface seemed as far removed from computing as possible were chosen. I interviewed two men who were mutual friends, both of whom were sculptors and who devoted most of their time to their art. Both were well-educated, quiet, gentle, humble men, who were also highly imaginative and capable artists. More detailed results of these interviews are presented in Appendix 2.

The results of these interviews revealed that there was little direct connection between the preferred activities and the past interests of the Dependents and the Sculptors. The Sculptors stated that they had never had any aptitude or interest in science, mathematics, electronics or new technology, the main interests of the Dependents. Nevertheless, the descriptions of their personalities, family background, the needs for and the pleasures gained from their activities did show the most extraordinary and remarkable similarity to those of the Dependents. For example, the Sculptors revealed similar problems with relationships and shyness from an early age. Their solitary hobbies had always been of paramount importance, they usually worked 'hands on' with little pre-planning, and were more interested in the 'process' of their work than in the end-product.

As the similarities seemed so striking, the Sculptors were also asked to complete the GEFT and the 16PF questionnaire to see if the resemblances were substantiated using the standardized tests. Even these results showed similarity with the Dependents in certain respects. On the GEFT the Sculptors were equally field-independent, and on the 16PF were shown to be as intelligent, reserved, imaginative, independent and self-sufficient as the Dependents. However, they were also more shy, introverted, tender-minded and humble than the Dependents and therefore even more socially inhibited so that they showed similar schizoid traits. The results of the two tests are outlined in Appendix 2.

The results from this sample of two may have been only coincidentally similar to those of the Dependents, but the interviews revealed that parenting methods may have once again been influential in the need to achieve and succeed at an early age in order to get attention. Much further research would be needed in order to establish whether the object-centred personality can be firmly attributed to early child-rearing habits, and whether it invariaby leads to the development of an intense interest in one particular activity, whether of an artistic or scientific bias, but the results from this study seem to suggest that this may be so. What started the Sculptors on the road to art and the manipulation of pens and brushes rather than the manipulation of machines was unclear; this may have been due simply to the availability of those items in the home and to parental encouragement or example, rather than to anything more profound.

The Sculptors' activities did not involve interaction with manufactured artefacts but with the manipulation and command of raw materials which they shaped into images which had no purpose other than to be aesthetically pleasing or intellectually challenging to the eye and emotions.

One can however see similarities between the construction and development of images and the creative writing of computer programs; the results demonstrated that the need for control over their media was just as important to the Sculptors as to the Dependents, and engendered the same feelings of power. However, as far as I am aware no-one has ever suggested that manipulation of a block of *lignum vitae* or clay is liable to lead to personality disorders or damage to social relationships, although such have been attributed to the Dependents' interaction with the computer; an interaction which was shown to differ little from that of the Sculptors with their materials.

From interviews with both the Dependents and the Sculptors I could only conclude that there was little difference between them; except that one mode of living was considered acceptable, even admirable, whilst the other was not. Both had discovered a manner of living which was in keeping with their personalities and interests; their activities and lifestyles may have been atypical, but they were fulfilled and happy. Others have adopted lifestyles which could be considered even more unusual than either those of the Dependents or the Sculptors, but they also have been found to be well-adapted to the needs of the individuals concerned, as a recent study of eccentricity by Weeks (1988) well illustrated.

Using similar publicity methods as used in this study, 130 people contacted Weeks stating that they considered themselves to be eccentric (74 men and 56 women), and Weeks and his co-researcher Kate Ward interviewed them all and gave them numerous psychological tests. The results make fascinating reading, no less so with respect to this research because many of their findings were similar to those observed within the Dependent group.

The eccentrics completed the 16PF questionnaire and were found to deviate significantly from the British population norms for two of the First Order factors, factor E (assertive) and factor M (imaginative). Such deviations were also observed within the Dependent group and both yielded similar mean scores; factor E—Eccentrics 7·5, Dependents 7·2, and factor M—Eccentrics 7·5, Dependents 7·4. The Dependents had also deviated on two other First Order factors, B (high intelligence) and Q2 (self-sufficiency) which were, perhaps surprisingly, not apparent for the eccentric population.

Although Weeks did not publish a profile of the eccentrics' 16PF sten scores with which comparisons could be made, he did extract the frequencies and percentages of those who had achieved extreme scores (a sten of either one or 10), reasoning that only 2·3 per cent of the general population would produce such results for each factor. This innovative exercise was carried out upon the Dependents' data and comparisons made between these two groups yielded 11 factors with scores greater than 5 per cent for either or both of the groups, see Table 11·1. (Only the

Table 11.1 Frequencies and percentages (in parentheses) of male Eccentrics and Dependents yielding 16PF sten scores of 1 or 10 (data from Weeks, 1988)

16PF personality factors		Eccentrics N=130 Frequency (percentage)	Dependents N=41 Frequency (percentage)
B+	High intelligence	10 (13·5)	19 (46)
Q2+	Self-sufficient	9 (9·5) (sic)	8 (19·5)
I−	Tough-minded	2 (3)	6 (15)
Q4+	Tense	6 (8)	4 (10)
C−	Affected by feelings	4 (6)	3 (7)
N−	Forthright	9 (12)	3 (7)
E+	Assertive	7 (9·5)	3 (7)
Q3−	Follows own urges	4 (6)	3 (7)
L+	Suspicious	7 (9·5)	2 (5)
M+	Imaginative	9 (12)	2 (5)
H−	Shy	4 (6)	2 (5)

data for males are given, as there were too few female Dependents to produce reliable comparisons).

This comparison showed that a larger percentage of the Dependents yielded extreme scores for factors B, Q2, I and Q4 (more intelligent, self-sufficient, tough-minded and tense) than the Eccentrics, but were less forthright, suspicious and imaginative. In all, 35 per cent of the Dependents and 46 per cent of the Eccentrics yielded two or more extreme scores, and neither group could therefore be considered typical of the general population, but even after extensive testing Weeks concluded that the majority of his sample could not be considered to have personality disorders. Of the measures taken, he did however conclude that the schizoid personality type was more apparent than any other, just as had been deduced from the Dependents' data.

There were numerous other similarities between the Eccentrics and the Dependents, both demographically, socially and psychologically. The majority of both groups were first-borns from the professional classes and had received tertiary levels of education. Not only did both groups show high levels of curiosity and originality in their thinking and activities, but the Eccentrics were as equally dedicated to their hobbies and interests as the Dependents. Both groups used the defence mechanisms of isolation and intellectualization, and from his data Weeks concluded that parental influences had led to the eccentrics' isolation from others and to their attraction to a world of ideas and objects. His description differs little from conclusions drawn in this research:

> 'They fell back on their own resources for amusement and solace, experimented with their environment and ideas, and generally extended themselves. Often this became intrinsically rewarding and increased the diversions of their activities and abilities, moving them further away from "normal" children.'

The quotations from the eccentrics, which showed equivalent levels of enthusiasm for their chosen subject areas as exhibited by the Dependents and the Sculptors, also exhibited similar levels of originality. Their ideas were nót necessarily rooted in the practical and the easily feasible, but went beyond into futuristic innovation and fantasy. They were not constrained by the *mores* of society but, as recognized by Weeks:

> 'Their personalities allow them to indulge freely, to the fullest, the whims of their vivid imaginations. . . . they ceaselessly assert their fundamental right to be what they want to be.'

From his results Weeks concluded, in accordance with the conclusions drawn from the Dependents' psychological data (Chapter 8), that the eccentrics were not neurotic but had achieved a level of self-actualisation rarely found. He suggested that they were happy and healthy, with a drive, sense of humour and abiding curiosity which enabled them to ignore the stresses of ordinary life in order to pursue their chosen paths without reference to convention?. He believed that 'certain types of deviant behaviour can be healthy, good and life-enhancing.'

A few of the Dependents had spontaneously described themselves as eccentric or felt they were considered eccentric by others. In general they felt unusual and different, and for many of the same reasons as found by Weeks seemed to have found the ideal path for the development of their personalities through the uninhibited, exploratory use of the computer. Although some used the computer as an escape from negative aspects of their lives, its use offered them a means of relaxation which removed many of their stresses, giving them contentment and satisfaction not previously experienced. From comparisons with Weeks' study one could conclude that computer dependency should be viewed as a therapeutic activity for such people. The Dependents appeared no more unusual than many other dedicated enthusiasts. They had developed interests which were in keeping with their psychological needs, strengths and limitations and which enabled them to function successfully, even if atypically.

Conclusions

From a study such as this it is difficult to propose firm, proven conclusions. The research had to be exploratory as so little of a factual nature was known about the syndrome of computer dependency, and for this reason some areas could only be examined somewhat rudimentarily. Not all of the Dependents who agreed to be interviewed could be, because of limitations upon the resources of time and money, and for the same reason none of the control group members were visited. The majority of the information therefore had to be obtained through the use of questionnaires and scales which, although providing ideal results from which statistical analyses could be carried out, did reduce much of the data to that which was easily measurable. There had been little guidance within the previous literature as to which areas to explore, since the authors seemed only to have described the activities undertaken at the keyboard or the effects arising from computer dependency. Any suggested causal relationship seemed to concentrate upon the features of the computer and rarely mentioned the individuals, other than to describe them as introverted. There was an obvious need to try to determine why this syndrome had actually occurred for some people who used computers.

That computer dependency occurred was established as far as was possible. The people who responded to the publicity and their partners all recognized and described the same syndrome as found in the literature. The Dependents devoted the vast majority of their spare time to computing, and for many most of what they did was only of intrinsic interest, done for intellectual stimulation and to learn about new technology. Most of the Dependents were not producing workable code, but were playing with the system hands-on and enjoying the resultant debugging. They differed from the control group of Owners in this respect but conformed to the image given to them by the press and the literature. What had not been anticipated was that there was more than one type of computer dependency. Although the majority of the interviewees were Explorers, as just described, some had graduated from this type of computing either to networking activities or to using the computer for work-related purposes. However, there was no difference between the groups for the effects that this had on others or for the time they devoted to their chosen activities; more important, irrespective of the group from which they came, both they and their spouses still considered them to be computer dependent.

The results from this study demonstrated that computer dependent people were not drawn from a cross-section of the population, nor were they typical socially or psychologically. These differences had been observed by some previous authors, but as most had made little attempt to try to understand why the syndrome had occurred one could only assume that their motives were not to offer assistance or personal advice if they felt it was required. However, most seemed to feel that such activities with a computer were detrimental, either directly to the individual or indirectly to computer systems, and as such should be prevented.

As the results from this research suggested that the Dependents had exhibited object-centredness and schizoid personality tendencies early in life, 'cures' offered in adulthood would have little positive effect even if one should wish to prevent the continuation of this condition. Merely altering behaviour patterns by weaning people away from computers could be very counter-productive. The computer was adopted because it offered the ideal interaction for those who had traditionally found great difficulties with personal relationships; computing was a fulfilment of needs, not an illness. For some it was a recognized escape from problems, but it was also seen more positively, as a substitute for something lacking in the outside world. Reducing the time spent on the computer would not solve the Dependents' problems and might in fact do more damage to the individual and the family than if the dependency continued. One man at least admitted that he had been cured from gambling by the advent of the computer in his life, and there were some suggestions of previous alcohol and drug 'abuse' by others. Few would suggest that the replacement of these would not be beneficial. Perhaps all who show personality traits similar to those exhibited by the Dependents need an object-centred, intellectual activity to give purpose, stability and control to their lives.

As suggested by Turkle (1984), we are perhaps all looking for an 'idealized person' with whom to share our lives, and some find it in a relationship with a computer. For the Dependents it was able to satisfy what we all hunger for, perfect acceptance and security. Others no doubt never find it at all and remain distressed and unhappy. If we are truly concerned about the Dependents and the thought that many more children will become like them as the computer becomes more widespread, we need to look at what is happening within the home during the pre-school years of the child, and especially within the middle-class home.

Although physical violence was occasionally mentioned as occurring within some families, the overriding picture which remained after the interviews was of people who, as children, had been well cared for in the physical sense but who had never been sure that they were loved. Follow-up interviews would have been necessary in order to explore this area more fully, but the distress created and the difficulty felt when describing their family backgrounds must be considered as highly significant for the development of the Dependents' subsequent interests. Whether rejected or neglected when young, they found acceptance through the performance of tasks and the achievement of goals and did not feel that they were loved unconditionally for being the child of the parents. The insecurities with which they were then left remained throughout their lives, only to be appeased by the manipulation of the trustworthy and reliable *inanimata.*

It is perhaps not surprising in a culture which appears not to like children that this situation should occur. No doubt many of the Dependents' own parents had received a similar upbringing and were therefore unable either to break out of the mould or to see any damage which could be caused by their methods. Such an upbringing was perhaps seen as advantageous, by ensuring that the children became independent early in life. It certainly appeared to have been successful in that respect;

however, not only did it encourage them to be independent from their parents but also from almost all other human beings. 'Dependency will out' should perhaps be the main conclusion drawn from this research. All need to feel secure and of use in this world, and the hopelessness which was felt by some of the Dependents early in life was replaced by a dependency upon objects which would not hurt them.

In a society which is prepared to spend millions of pounds upon child abuse after the event has taken place, but offers so little in the way of marriage guidance, let alone lessons in parenthood and marriage while at school, it seems unlikely that things will change. The noble district nurse would notice nothing abnormal about the family and would therefore never suggest that help and advice should be given. It is perhaps the middle-classs mother who is less likely to take employment while her children are young, thereby keeping her children with her and out of nursery schools, and less likely to have an extended family around her to offer additional carers to the child, other than the occasional au pair. If so, such situations will probably be perpetuated.

The family members would probably be unaware of any difficulties, and it may be the child's first teacher who notices a reclusiveness, an avoidance of contact with other children, and an aloofness from adult caring. By this stage it may be too late. The size of classes may inhibit the teacher from giving sufficient personal attention to the child, and a child who will quite happily play alone will not demand the same attention as would the more boisterous and gregarious children. The child would be allowed to continue in its object-centred activities, and will be shunned by other children if showing interest in the more academic activities. This will be perpetuated throughout life, as even within secondary and tertiary education the studious tend to be derided and have to pretend that they do as little work as possible if they are to be acceptable to others.

Thus one can only advise all parents, who must desire the best for their children, to make positive efforts to show overt, physical signs of affection towards them when they are very young. It is not sufficient to offer them the best of material goods and the best schooling; children's needs are more basic if they are to grow up to be secure, happy and well-rounded adults.

Sex differences are also apparent in this issue, as the large proportion of male Dependents indicated. Cursory observation suggests that it is the male who is more likely to concentrate upon one particular subject or activity in order to perfect it, while females tend to be more eclectic in their interests and are more likely to be content to be the 'jack-of-all-trades' and know a little about a lot of things. This can be seen reflected in the contents of magazines aimed either at the male or the female markets. The 'male' magazines will, for example, concentrate almost exclusively upon one topic, whether it be fishing, football, cars, motorbikes, chess, pornography, or of course computers; invariably they centre upon one object-centred activity. Women's magazines are less exclusive, and even the few which may appear to centre upon one topic (such as fitness) will probably all contain articles about food, clothes, hair, exercise, kitchens and houses in general, as well as stories about people, and personal relationships and their attendant problems.

Some may suggest that women have no time to specialize in one particular subject due to family commitments, but the schoolgirl and the single women with similar diversity show that this is not the reason. Our early socialization must form one of the major causes, leading the majority of boys to exclusivity and the majority of girls to diversity and interactions with people. In babyhood it tends to be the female child who is praised just for being; being pretty, clean, smiling, quiet, etc, and for expressing emotions. Malatesta (1985) demonstrated that mothers would often ignore the emotional expressions of boys but encourage and respond favourably to those of girls. Boys on the other hand are encouraged to be physically active and to 'do things'.

These tendencies continue. While at school boys are encouraged to struggle with the difficult 'masculine' subjects, moulding them to the desired image, while girls are allowed to opt out (see Chapter 5). But those who have repressed their emotions through such early conditioning, usually males, are not given extra lessons in feeling and caring, and this research has demonstrated that they are often the ones to be actively encouraged to attain inaminate goals. Even in these days of sexual equality, girls are allowed to explore their psyches and emotions at will, and are still eased into the caring professions to exploit their strengths, while often allowing the mathematical and scientific sides of their characters to remain undeveloped. Both sexes are losing a great deal, and it should be of little surprise that communication between them is often unsatisfactory. In spite of the worthy work of the Equal Opportunities Commission, among others, there is still a long way to go before boys and girls are seriously encouraged to develop not only their strengths but their weaknesses; weaknesses which have been instilled since they were infants, because one sex is more highly favoured if it expresses emotions and an interest in others and the other if it does not.

The childhood difficulties experienced by the Dependents were often ameliorated to a great extent by marriage, which brought them the security they so desired. In some cases, however, as their problems diminished so the problems of their partners increased. The impression was gained that some of the wives of the Dependents had little knowledge of their partners' backgrounds, or an understanding of the effect it could have upon them. There were also suggestions that some of the wives' backgrounds were not entirely free from trauma and abuse of some sort, and for this reason they may have turned to a partner who so obviously needed them, perpetuating any problems. The need for wives to love those who seem unable to love in return is discussed by Norwood (1986), mainly with reference to wives of alcoholics. She describes the early experiences of such women as being:

> 'either very lonely and isolated, or rejected, or overburdened with inappropriately heavy responsibilities and so became overly nurturing and self-sacrificing . . . Inevitably, she will involve herself with a man who is irresponsible in at least some important areas of his life because he clearly needs her help, her nurturing, and her control. Then begins her struggle to try to change him through the power of persuasion of her love.'

This was shown to be true within some of the Dependents' relationships, as was the fact that the wives were usually unable to change the behaviour of their husbands. Their lives depended upon the care of their man and upon his control. One wife, a psychologist, was well aware of the very traumatized childhood of her husband and was prepared to indulge his needs to help ease the pain, the cause of which the husband seemed unable to recognize. However, although she seemed to have managed to convince her husband of her love for him, to the extent that his dependency upon her was almost total, she was not able fully to replace his long established interests by her own needs.

How can such marriages be helped? To summarize the advice given by Norwood, she suggests that wives have to admit that they are helpless to stop the dependency and need to start to take charge of their own lives, by focusing upon themselves and their feelings using the support of others. This may help the woman, but as it is not a shared experience of the couple it may not save the relationship. From the basis of this research, additional assistance could also be suggested. As the relationships which showed the greatest problems were not always those where the wives were reluctant to use computers, perhaps one could suggest that the very medium at the centre of their difficulties could be utilized to help them. One feels that there would be scope for some computer software which would enable the partners to explore each others' needs and personalities in more depth than previously.

Someone commented that they had observed two 'junkies' communicating with one another via the networks, whilst sitting only a few feet apart. To those with good social skills this may seem bizarre, but this type of interaction successfully removes the non-verbal cues from conversation which can cause distraction and distress to some; the very things which give life to conversation to most people can be torment to others who cannot read the signs accurately. Exasperation, boredom, anger and frustration are not readily apparent on the networks, even if felt. A question mark appearing on the screen is a helpful indication that something has not been entirely understood and is far less distressing than to be told 'I don't know what you are talking about' or 'Forget it, it doesn't matter'. On the networks, one is not interrupted and nor are one's sentences completed inaccurately by the other person, for example, and this method of interaction seems perfectly to suit the Dependents' logical, precise modes of thinking and reasoning.

Face-to-face interaction was often difficult for the Dependents, especially when talking about their feelings and backgrounds. Perhaps they were rarely understood by their partners because they so rarely explained themselves to others. A program which allowed each person to question the other and to express their opinions somewhat privately, without the problems of non-verbal communication and outward emotion so apparent within the couple's interactions, might prove of great benefit to both. Neither the Dependents nor their partners were insensitive creatures who wished to hurt the other, and as both were invariably committed to their relationships such a program, using a medium considered unthreatening to the Dependents, may have some success.

It was not difficult to empathize with both parties during the interviews but their lack of communication, at a level which both could appreciate and understand, inhibited successful adaptations being made to their behaviours. I found it somewhat disturbing to be told by some of the Dependents that I now knew more about their histories and family backgrounds than did their partners. I obviously offered no threat, was impartial and non-judgmental, and perhaps the computer could act as a similar intermediary. One felt that if both had been able to share their needs, interests and insecurities with one another then greater understanding would have been achieved.

Males in particular will need help to express themselves, and from an early age if they are to realize their self-worth as human beings, not just as the performers of tasks. The interviewed men often felt totally innocent of the charges of neglect laid before them by their partners; they were at home, not in the pub, were stimulating their intellects and not wasting time, and had often through their interests increased the family income considerably. What more did women want, and why? The use of the computer as a device for communication may help them to discover the answers, and help the wives to understand their partners' difficulties.

From the results obtained, it did seem somewhat strange to discover that this group had been singled out for special attention, to an extent which had inspired this research. The interviews revealed the Dependents on the whole to be a group of middle-class, young, intelligent and well-educated people who were happy with their lot, and invariably very successful if this may be measured by the status of their jobs. It was therefore somewhat difficult to see how they were not merely conforming to the images presented as desirable today.

The philosophies of the current age positively cultivate the fulfilment of one's own ends regardless of others; the welfare state in Britain appears to be in the process of being dismantled and it is 'each man for himself'. The desire for success, prestige, status, money, and to be at the forefront of technology are all encouraged, within our schools and universities, by the press and the government. Interdependence is not fashionable; there are no common causes, and a sense of belonging to a society is no longer encouraged; science is worshipped, art relegated. Western culture is now object- and task-centred; the nurturing, caring society seems all but to have disappeared and the present political influences are apparent even within the student population, traditionally the site of radicalism and social reform. Self-need and individuality are fostered at the expense of the group. Life becomes work-centred and many people have no time for others, not even their families. Personal fulfilment has become paramount; there is no longer the need to make an effort to belong, or to be concerned about the welfare of others. This could be described as a 'schizoid' era, with the Dependents merely conforming to this image.

In the society within which we live, especially for the employed middle-classes, basic needs are taken care of; there are no wars to fight nor national goals to strive for. It is not difficult to clothe, feed and house oneself, and to achieve self-actualization and fulfilment individuals have to create their own personal challenges and excitement. This the Dependents had done very successfully, but they differed from others in the fact that they were not necessarily achieving anything tangible from their labours. Most were not getting paid for the many hours they devoted to their activities at home; it was done for sheer pleasure and enjoyment, and perhaps this marked them out as different.

As a conclusion to this study, I offer therefore no list of predisposing factors one should look for in a child so that computer dependency may be recognized and prevented; after all the child may instead grow up to be a famous sculptor, racing driver or high-technology innovator. What I would like to suggest is that perhaps our society needs people such as this; perhaps a diversity of extroverts and introverts is healthy and to the benefit of us all. Perhaps we need 'thinkers' and 'doers' if we are to make progress. From the work of Rosner and Abt (1970), when describing the creative experience from interviews with some of the most famous scientists of the day, one could easily see that backgrounds similar to the Dependents, full of shyness and isolation, may lead to the development of wondrous inventions and ideas. Why then, one might ask, were the Dependents not achieving such heights? Give them a little more time and perhaps they will.

Appendix 1
Case studies

This appendix contains a series of case studies, compiled from the interview data. All the names have been altered to ensure anonymity, and each case study has been labelled according to the main computing activity preferred by the interviewee (see Chapter 9).

As well as being chosen to represent the three types of Dependents, the case studies illustrate the most extreme cases found and the extent of the severity of the problems experienced by some of the Dependents. The people described in case studies A and B are both Networkers; those whose main use of the computer was via a modem to contact other databases and individuals, to play games or for hacking. Those described in case studies C and D are both Workers; their computing consisted mainly of programming with a practical and useful end-product in mind and in using commercial software. Finally those described in case studies E, F, G and H are all Explorers; whose computing was exploratory and self-educational in nature and who rarely produced extrinsically useful end-products from their programming.

Case study A

'Networker'
Male, early forties, married, no children
Time spent computing per week: Home–28 hours, Work–40 hours

Mike's family background was 'happy but very quiet, a house with very few visitors'. He described his mother as strong, stronger than his father, and not tremendously caring, but stated that he 'didn't suffer'. His father, though quieter and weaker than his mother, was the disciplinarian.

His primary school days were rather difficult; he felt set apart because his mother was a teacher at his school. Although he described himself as more shy and serious than average, tending to lead a rather solitary life, he managed to cope as he was also 'rather tough and a fighter'. His secondary schooling was more enjoyable; he took on the responsibilities of prefect and rugby captain and went through no rebellious periods. He disliked languages but enjoyed mathematics, science and sports, and found school work easy. He had been very interested in stamp-collecting when young, a hobby later to be replaced by electronics and various types of puzzles, becoming 'completely obsessed by the *Rubic Cube*' in its heyday. Academically successful at school, he went on to gain a science-based B.Sc. and Ph.D.

As an attractive man, Mike had experienced no serious problems making friends or finding girlfriends, although he felt himself to be a late-starter having his first girlfriend at the age of 18, a factor he attributed to his quiet, rural upbringing. He admitted that he was always more seduced that seducer and that girls had 'thrown themselves' at him in the past. Mike said he 'loved women' and married at 23; a marriage he described as happy, seeing his wife's abilities as complementary to his own.

Mike described himself thus: 'I'm an introvert trying to be an extrovert. Although I was quieter and shyer than average I think I'm superior–you always do if you have a Ph.D.—I tended to lean on the Ph.D. a bit at work. I wish I was a better communicator—that's probably why I like computers. I don't have any small-talk and I stammer when in a group, but not individually like this. At work I'm respected—they look up to me. Apart from being a computer eccentric I think I make friends fairly easily'. This last statement was later contradicted when he said that he found making friends more difficult as he got older, and although he had 'picked up some new computer friends on the networks' whom he occasionally met, because of computing he rarely spent any time socializing.

He has spent all his working life with one company, steadily rising through the ranks to managerial level and aims to be 'on the executive one day'. He initiated computing in the company, designed a new post to take responsibility for it and promptly filled it. He loved his job, which gave him great autonomy and responsibility, but his interest in computing had caused some problems at work. He admitted to trying to delegate many of his tasks so that he could 'be left 100% on the computer', and felt the management aspects of the job were suffering as a consequence.

Mike first used a computer at university and since then had always seen his future with new technology. He bought his first microcomputer in 1981 in order 'to find out about computers—for enjoyment'. He stated that he was immediately 'hooked', and had been ever since. Initially he spent most of his time copying and modifying the programs of others, 'a lot with no purpose. While learning BASIC I wrote a program to demonstrate the use of the computer. It was the only program I wrote which had an end-product'. He described his programs as badly written and undocumented—'Debugging is the enjoyment of programming, I enjoy changing and adapting things.'

He no longer has any interest in programming while at home, and now spends most of his time playing games and networking. Before midnight his time is spent on the networks or copying programs, after midnight he plays *MUD*. This was his chief passion, and he tried to play every night. Although admitting that computing took up too much of his time he did not wish to stop, 'I would devote every minute of my time to it'. He loved all types of adventure games for the intellectual challenge of solving the mysteries and admitted that he had taken a week's holiday from work to get to 'Wizard status' (the highest) on *MUD*. 'I get a real high, an adrenalin high. It scares you, but you begin to live on the challenge. With *MUD* you strive to be a Wizard—it's a tremendous high. It's the intellectual challenge which I find so absorbing—more so than humans.'

Because of his expertise in this area, he became an adventure game advisor for one of the networks. 'I have to be the best—I'm challenged all the time. I like the competitive situation, but I have to be the best'. He sees his computing as a hobby, a means of enjoyment. 'Throughout your working life you are given challenges, but with a hobby you can tailor it to your own ends'.

He admitted he would spend even more time than he did on the computer (an average of 28 hours per week at home) if he was single, and stated that his wife did at times get very annoyed with the time he spent computing, a situation which led to arguments. A major change which had occurred within their marriage was that they no longer went to bed at the same time. She had usually slept for many hours by the time he finished computing, a situation which caused her some distress.

His wife however did appear very tolerant of his computing, and seemed to find it preferable to other activities in which he could have indulged. 'Occasionally we fall out. I blow my top occasionally and we do argue. He is aware and conscious of how I feel. I use a computer at work and I can see the fascination for him. It does make him happy, and at least I know where he is.'

Case study B

'Networker'

Male, late thirties, married, two children

Time spent computing per week: Home–25 hours, Work–45 hours

Richard grew up as a 'bright lad in a poor, working class home'. He found it difficult to describe his parents, but stated that they were undemonstrative, his mother 'bright, but uneducated', his father 'arrogant, self-important and demanding'. He never felt close to either parent—

'I've very little interest in my family. I only visit them out of pseudo-middle class duty.'

His school years were miserable. 'I hated school. I liked chemistry and hated all the rest. I was brighter than the rest. I always felt totally alone in the world; always a loner. I like to think it was because I was moved up a year at school when I was eight. It tends to isolate children, I then had to stay for two years in the top class. Who can analyze it—I've never been gregarious? I was more clever, I had the imaginative mathematic ability, sideways reasoning that they lacked.'

Problems became so severe at school that he was suspended for truancy, and then attempted suicide. 'Life didn't seem worth carrying on with in cold, stark analysis. I'd just been suspended and a girlfriend had just left me. It was not a very appealing outlook. I turned the gas taps on and I woke up in hospital. I caused my parents a lot of grief. I didn't try again, but life is not important, not something to hang on to.' He returned to school, but after a road accident was hospitalized for three months. His examinations suffered and it was some years before he returned to formal full-time education.

He attended university in his early twenties, found the work easy and spent his spare time 'on sex and dope'. He is now very happily married with two children. His wife seemed to provide the greatest support in his life, and even the hours spent computing were said not to upset her. 'I think she may feel guilty about deserting me when she goes to bed. She needs sleep a lot and I don't. She knows I feel alone in the world.'

Although having no career plans and stating that he did not enjoy work, Richard became a science teacher. In 1979 his department acquired a microcomputer which no-one used. He took it home for the summer holidays and became 'hooked immediately', spending most of this introductory time playing and programming games. As his interest in computers grew, he initiated the development of a computer studies department within his school and became its head.

He devoted nine hours a day to the computer department, from 9 until 6, running clubs during his lunch-hours and after school, but rarely becoming involved in other school activities. 'In role I'm okay, i.e., being a teacher, but when I have no role and have to be myself I prefer to be alone. In school the kids think I'm great, but I'm not a great one for relationships with other teachers, they are not very bright'. He enjoyed the 'ego trip' of being a teacher, but would have preferred to spend all his time on his home computer if he was not in need of 'money to pay the mortgage'.

Richard's home computing now centred around networking, especially playing *MUD* which he considered to be the ideal outlet for his talents. He tried to confine *MUD* to the weekends, when access to the Essex computer was less restricted, and would often play for nine hours at a stretch on Friday and Saturday nights. Other activities involved 'zapping aliens' on other computer games, indulging in software piracy (he was involved with 'three major pirating circles') and using the networks to contact others and access software. Although he 'spent every spare penny on computing' this was confined to the purchase of hardware and blank discs. Of his £10 000 worth of software, only £15 of this was said to have been legally purchased. He had a large collection of utilities and communications software, but most were games. He often only played a game once, preferring to extend the programs having broken the software codes, although he did play *Elite* (an adventure game) solidly for a month.

Only five per cent of his time at home was spent on programming, which was always of the exploratory kind. 'I'm not a talented programmer, it's easier to hack others around. I'm not conventional—flow charts, ugh!' He had broken into six university networks and found hacking 'wonderful—it's great fun—I enjoy it. I get a big kick out of breaking someone's protection software—that's just kicking back at society'.

Richard did not view the world through rose-coloured glasses, and freely admitted that he used computing as an escape. 'I do it because I have very little else to do, watching television is a waste of time. Computing is used as an escape from reality. It's for idle time, it makes the hours flash by and relieves the boredom of life. I'm depressive, life is boring. I see no sense, purpose or reason in life. I have no great aspirations, there is nowhere I want to go or be. I want to be comfortable. I sell my labour for money. Computing is an escape, I'll watch television to escape also. I watch a load of old rubbish. It's escapism, so is *MUD* and science fiction. Everything is an escape from reality, like smoking dope, it takes the sharp corners off. I'm dissatisfied with myself. I have no sense of purpose. I lack aims and objectives and I don't really approve of me. I'm verging on the manic depressive, it would be very easy to go over the top.

'I don't socialize, I didn't before. I used to make the effort, but don't bother now. I don't really enjoy social activities. Relationships are painful, they hurt. You get carried away with it. There are no other relationships that sexual ones—there is nothing else. Religion and politics were invented to come to terms with that. University was fun—little peaks of excitement—all were ephemeral. It all comes down again to the gloom of existence. There are some nice highs that I will remember. The unhappy times are general, I don't look for something to happen, I sit down gloomily until something happens.

'There is no real me—it's all only a game. I can play at being an extrovert or an introvert. I'm somewhat obsessed with dope, sex and computing in that order. There really isn't anything else. I only do things if I'm interested. I don't know whether I'm insecure and just pretending to be confident. I can talk to you, I couldn't do this on the 'phone or with a a man. I don't like getting old, 30 was a real crisis, 40 will be dreadful.'

Case study C

'Worker'
Male, early thirties, married, no children
Hours spent computing per week: Home–15 hours, Work–30 hours

David, although not a difficult person to interview, gave responses which were brief and to the point and seemed unused to contemplating or discussing personal details of himself or his family. From a middle-class family, he was 'not close to either' parent, describing them both as distant, but strong and dominant. He attended boarding school which 'weren't the greatest days' of his life, but was academically very successful, gaining his ordinary level and advanced level examinations two years earlier than normal. He left school at 16, having been a prefect and house captain, and joined a large computer company for a year, before going to university where he read electronic engineering.

He first used a computer whilst still at school, having access to a university's mainframe nearby, and quickly became very involved. This was the stimulus from which his interest in electronics and computers sprang. He purchased his first microcomputer in 1975, a kit from the USA. After assembling it he found there was little to be achieved until he had built an operating system for it, during which time he learnt a great deal about electronics. He had, since then, owned six different computers and was familiar with eleven programming languages, some of which he needed for work (he had helped design one of them), and others which he learnt for his own pleasure. The majority of his computing activities were work-related, even when undertaken at home, and he had been known to spend over 30 hours programming without a break. He considered his programs, which he said rarely needed debugging, to be well-written and some had been used as example texts at work.

Describing himself as shy, unemotional and quite introverted, David only had few friends while at school, preferring to spend his time in solitary activities. He did not find it easy to mix with females, a factor he blamed upon his male-dominated degree course, and he did not have a girlfriend until he had graduated. He returned to work for his original company where he now holds a senior post, and met his wife over one of the internal computer networks. Marriage, to an equally successful and attractive woman who herself used computers, had brought him great happiness, although his job still provided his greatest fulfilment in life. David now tended only to mix with people who worked with computers, and recognized that his computing had cut him off from 'the very humanist people'.

David described himself as a perfectionist; mature, ambitious, rational and intelligent. He considered himself ideally suited to computing which had provided him with the ideal job. 'I'm a fairly tidy person—the needed skills are my skills, the ability to think through a problem, being logical. Also computers have the ability of being continuously challenging. However much you know there is always more . . It's nice to get it to do what you want it to do. It's nice to get something concrete and physical out . . . A tool, is the best description (of a computer)—it enables me to achieve much more than I can on my own. It's a challenge.'

His job had provided him with an excellent salary, he lived in a beautiful house in idyllic surroundings, and he enjoyed the conditions and resources which had been made available to him through his work. David seemed a very contented man, who although working long and hard did not appear stressed and had realized his ambitions.

Case study D

'Worker'

Male, early thirties, married, two children

Time spent computing per week: Home–25 hours, Work–10 hours

Born abroad, Steven was sent to England for his secondary and tertiary education. Although a slow starter academically he worked hard, achieving five advanced level examinations and a good business degree, also studying accountancy 'at home, to better myself'. He went into management after university, changed jobs frequently, each time rising in position, and is now managing director of a company with branches abroad. Affluent, he owns a magnificent house near London and another in the Far East where he also works, with two computers in each. Although they all used to live abroad, his family now remains in England while he travels, sometimes being away for long periods of time.

He was the eldest son of four children and did not enjoy his childhood. His father he described as: 'strong-willed, a patriarch, aloof, estranged from me, totally cold. He never carried me, kissed me or talked to me. He was very well off, but mean. He made us live like poor people, he lived very well. He would discipline very violently. He was violent to my mother and had other women. She was powerless, she suffered and put up with it. She was kind and loving, always on our side. She was a kind disciplinarian, but neurotic and depressive. They did not have a good relationship. They separated after I left home; she left him. The children help to support her.'

Steven described himself as very shy when he was young. 'I'm still more of an introvert that an extrovert. I used to be very nervous. As an adult I went on speaking skills courses, they helped, I needed them for my job.' During his teenage years he went through a period of 'controlled rebellion, against conventions, rules and regulations, which were what my father stood for—in limited ways, in small ways, just not conforming.'

His ambitions centred upon wanting to escape from his father's influence and support, and although having no specific plans, he always wanted to achieve and obtain a well-paid, prestigious post. 'I'm a self-driven man, a little bit like a machine. I think logically and recognize that others don't. I strive to be logical. Illogicality is a human failing which needs correcting. I drive myself very hard, I always have. I direct my life, make deliberate choices'.

He stated that he had always been very obsessive about his hobbies, dedicating a year or two to each, and needed to be always doing something useful and stimulating. His reading matter was non-fictional 'technical computers, business, science, banking and finance'. His social life had been female-centred, and he had had a number of steady relationships before he married. He said he enjoyed social activities, especially playing chess and bridge, but now immersed himself in his work and computing. Although stating that his level of output was essential in order to run two homes, he greatly enjoyed his work, especially the challenges and the responsibilities involved.

He had bought his first microcomputer in 1981, having had very little experience with computers, dividing his introductory time: '50 : 50 between exploring the system and writing specific programs. I thought I'd be good, but I have done more than that, I have a talent for it'. Although his company had a mainframe, he introduced microcomputers into the branch where he was a manager. 'It improved sales and my salary. The company has many branches worldwide and we were the first to have a computer—very advantageous at very little cost. Computing gave me a lot of prestige, success and advanced salary when I introduced them at work. It improved everything. I also have a business selling them abroad. People come to me with their problems to the group (he set up a users club abroad), they come to me with hardware and software problems.'

The vast majority of his computing was now work-specific, although he was unable to do much computing himself at work because of his managerial responsibilities. Nevertheless he still spent most of his spare time computing and had written many programs for work which had been adopted by the company, 'Most are mine—an accounting program for the office, the whole system—mailing lists, analysing marketing data, sales and quotations data, household expenses—they are all in use. Most of my work now has a useful purpose, for work or for others.' In spite of the usefulness of most of his programs, Steven was one of the few

Workers who still programmed hands-on and rarely documented his code, and was in the process of writing a business-based game which he hoped to sell. He had also just acquired a modem and was going to try his hand at hacking.

He now believed himself to be 'slightly less hooked. I was especially hooked at the steepest learning curve in the first two or three years. I got a little less hooked when I felt I had mastered it, it had satisfied my own needs.' This was nevertheless disputed by his wife who had seen no alteration to his behaviour, in spite of his protestations that 'as a family man I very consciously try very hard to be loving, and spend as much time as I can with the children.' Steven's computing had created serious problems within the marriage, to the point when his wife had returned to live in England with the children. This was effectively a separation, as most of his time was spent abroad only seeing his family during the vacations.

His wife was very hostile towards his computing but had made some attempts to understand his fascination, to the point of studying programming herself. This had tended to only make matters worse. She proved to be a far more efficient programmer than her husband but still could see nothing of intrinisic interest in the machine. 'After a year she went on a year's Cobol course to retaliate. She was good, but she never wanted to do more than the job in hand. She only uses the computer for word-processing and accounts, it's just a tool.' Not so for Steven.

'I spend all my spare time computing when alone. It's an excellent companion and it's of use. It gives me power—it's a machine that I can run, I can cleverly control it. It's your servant, it gives a psychological buzz and I lose track of time. It's a way of relaxing. It's a powerful tool—micros will become powerful tools. It's a machine—it's the product that's important . . . it gives me a feeling of power.

'I have to have time to myself, to be alone. I need to be quiet and relaxed, but the computer would relax me if not. I feel it's been something that's enabled me to continue exercising my mental faculties—without that type of exercise I would have made them more redundant. I always thrived on learning new things, passing new exams. As you get older you feel less able of learning new things. Computing has made me realize that I can still learn.'

He believed that his computing was sensibly controlled: 'I don't find it difficult to stop, I do it when I have nothing better to do, i.e. instead of watching television. In the family it may have alienated me somewhat, it's possible, but I'm not aware of it.' His wife was, however, and stated that she and the children had suffered a great deal since computing had come to dominate his time. There were serious communication problems between Steven and his wife, perhaps not aided by the comparisons that he made between his interaction with the computer and with people.

'I recognize a distinct difference between a human being and a logical machine. A totally logical person would be a machine. You cannot expect humans to be logical all the time. Computer interaction is much less stressful—it's logical and I come across a lot of illogical things managing people, it is stressful.'

Case study E

'Explorer'
Male, early forties, married, no children
Time spent computing per week: Home–40 hours, Work–25 hours

His father, a doctor of law, died when George was two years old, and he was brought up alone by his mother. 'My life wasn't happy. My father died and I had no substitute father-figure. My mother was terribly possessive, but she couldn't mould me to her ideas. I was totally disenchanted with her until her death. She was very trivial, always nagging me about being untidy and noisy. She couldn't care less about me, but wanted me to do well. Our relationship got worse as I got older. I would like to know what my father found attractive in her.'

'I was terribly lonely as a child. I always felt special and rather odd.' He hated his school days and felt that 'the teachers were hopeless. I never wanted to learn what everyone else was learning. I was very bad at most subjects at school. I wasn't popular. I always knew more than the teachers—I have a photographic memory. Most of my education happened outside school—self-taught from books. I always had one or two subject areas which I'm devoted to since childhood. I never did homework, never stopped for meals.' He reported that he taught himself to read at the age of six and read avidly, chiefly from encyclopaedias.

George found it difficult to talk about personal matters—'I do not see the point in questions about my family, what relevance are they?'—but his wife added more details about his childhood. 'He still says little about his mother, he hated her. They quarrelled the whole time. Even when we used to go and see her once a year, they always behaved together as if he was three. She wanted a conventional little boy working in a department store. It was a poor parental match. His grandfather lived with them, but died when he was five—he was very special. George used to drink and they used to physically fight, throw things. She was not at all indulgent with him—couldn't care less about him, she did not understand him. She was not encouraging, she reduced him the whole time, she always compared him unfavourably with his classmates. She never bought him any toys or even coloured pencils. His mother promised treats and then backed out, he never looks forward to anything. She never let him out in case he mixed with bad company. She bought him clothes, never toys. She used to hide cakes from him, they were for visitors. I make him cakes to spoil him—he'd eat them until he was sick. I indulged his need for special treats . . . Of course he's neurotic, to withdraw into machines away from humans is neurotic.'

George went to university in his home town, gained a doctorate in mathematics and remained there to teach. In spite of his poor relationship with his mother he lived at home until he was 27. Although he threatened to leave many times, he did not as 'there was nowhere else to go'. At university he only had a few academic friends and spent most of his time drinking. 'Lots of drinking—I was an alcoholic for years—more drunk than sober most of the time.'

He moved to another university, 'it was either that or murdering my mother', and met his wife. He had not started 'dating' women until he was 23 and had only two relationships, both with women much older than himself. His wife was his landlady, 'I became his mother, the ideal foster mother. I didn't expect anything other than an affair but he insisted on marriage. I felt sorry for him—I got to know him well. He found anyone difficult to talk to who was not on his wavelength—he didn't like giggly girls.'

George became animated when talking about his computing. He had taught himself to program during his first teaching post, and became a lecturer in computer science. 'I was hooked then. I thought they were marvellous tools. I got more and more involved with computer languages as such, rather than using it as a tool.' He purchased a microcomputer because he 'wanted to be informed—because the students knew more and I knew nothing. I just wanted to know about micros. I write programs simply to explore the system in the main . . . Self-education is the most appropriate description of what I do, because I just want to play around with it. Now I have a good reputation as a BBC expert—I'm well-informed. Others don't know as much—even *Acorn* (the manufacturers).'

It's so fantastic now (while talking about a particular editor he was writing). After two years it can do more than two or three commercial chips. I've forgotten most of the commands. It can run from sideways RAM and will be on ROM soon. I might get interested in a particular program which I might modify, or discover bugs in it—I have to put that right. It's absolutely futile, but I enjoy wasting my time on it. I break codes and things. Some have three or four protections, all mixed up—it takes ages to crack them—they make it more time-consuming. I buy cassettes, put them on to disks, make menus, improve them. It's all for the sake of perfection—so I can say I have all the programs from one magazine from the year dot. Sometimes I do scientific things even—Pascal modifiers, etc. I don't care about the end-product—they shouldn't need to have any useful end-products so long as I'm happy, that's the important thing. I'm a very odd character—what I enjoy most is improving other people's programs—so you can be immortal—'smart bombs'—I improve everything, graphics, sound, speed etc.'

George's computing was a very personal affair, and rarely shared with others. Although mainly programming, he did use the computer networks, but merely for accessing data. 'I'm not into making contacts—my computing is very private.' George stated that he was a 'hands-on' programmer and did not pre-plan his work, leaving him with programs which even he found difficult to read, a situation which did not seem to concern him. He used his computer every evening, often until about 3 a.m.

Even after many years lecturing he had no interest in applying for higher posts. His wife reported—'He doesn't want promotion. He's a true scholar, he wants to be left alone to work. He's at the top of the tree without any administrative responsibilities and distraction, he's not interested financially'. He loved the comparative freedom of his job. 'If it's raining I don't go in, I work at home. There are not many jobs like that.' To him work is play. 'I use the two words synonymously and have done for years.' His life was described as relatively contented, which he attributed to 'keeping myself occupied—not thinking about life in general'.

He explained his computer dependency thus. 'You get amazing feedback from the computer, it makes it fascinating, the instantaneous feedback makes it exciting. It's a type of communication you can compare with a chemist's formula. The feedback is vital—it's narcissistic—you get feedback from yourself. It's so gratifying—even when I can't beat the program—I've lost—but I've still won because I wrote it. I need to find out what its limitations are, that's a respectable use for a home computer. I have complete control over it—what you can do and not do. It's only limited by memory and time.'

He and his wife lived a very reclusive life because of his reluctance to mix with others socially. 'It's not at the top of my priorities. I've never had any intimate friends. My wife knows a lot about me and my past. I'm not deliberately secretive but it doesn't come naturally for me to talk.' To quote his wife—'He's only friendly with academics, only with the oddest. One is a microcomputer addict, he's another only child and unmarried. All his friends have a complete absence of private life. If he's bored during a social evening he'll pick up a newspaper and ignore everyone or start writing.'

George admitted that much of his computing was done as an escape from people. 'I don't understand people normally. I cannot look through them. I get on well with my students, but I'm not a person who understands them. If I watch a play I don't get the point of it. People are not as honest and open, they are not logical. Maths is either true or false—with people it doesn't apply. I have no ambitions to understand people this way.'

Case study F

'Explorer'

Male, late forties, married, no children

Time spent computing per week: Home–25 hours, Work–30 hours

Frank, the second son of middle-class, professional parents, experienced a great deal of trauma in his early life. (While discussing his background he became extremely distressed, but nevertheless wished to continue with the interview.) He described his childhood as isolated and lonely and in addition had little in common with his elder brother with whom he felt he was in constant competition (whether this was for his parents' attention was not established). The elder brother's dyslexia allowed the younger to excel academically, nevertheless Frank always believed that his father was disappointed in him for not being more physically competent. Six months of missed early schooling, because of illness, left him feeling estranged and unable to relate to his peers. He hated his school years, describing his later national service as more enjoyable as it was 'less tough than an English public school!'

His parents had an unhappy marriage and his mother, a manic-depressive, committed suicide when he was 11 years old. 'I never cried when she died. I had no emotions from that age until I met my wife.' His father became an alcoholic shortly after her death, and after a number of years of drinking and debt also attempted suicide, but failed. Frank's responses on the attitude scales showed his mother as distant and not loving nor possessive, while also being anxious, changeable, neurotic, sensitive and frustrated. Conversely his father was described as loving and out-going, but also as weak, emotional and anxious. In order to cope with stress, Frank was prescribed phenobarbitone while in the sixth form, to which he stated he became addicted, being 'forced to take them daily' by the doctor.

Social interaction was always difficult for Frank. 'I was extremely shy and introverted as a child. I never learnt what people expected in human interaction. People would say I was extremely difficult . . . I've a rigid personality, but have sufficient self-awareness not to be too anti-social. I'm introverted, I was very neurotic. I built up an intellectual way of life which allowed me to function as a useful member of society. It didn't make me happy, but it didn't make me miserable. I was frightened of people. I should like to love everybody—I have trained myself to stop disliking people. I would love to be worthy of love, but I know there's no such thing.

'I knew I was abnormal. Other people wanted to go to parties, to laugh and joke, and I didn't. In seven years at university I only went to four or five dinner parties. Since I met my wife we do meet people for drinks and now I can enjoy it, before it would horrify me except on an impersonal level as at work.'

Gaining a B.Sc. and two higher degrees, Frank has had a successful career centred around computers, being interested both in the hardware and software. His dependency upon computers had lasted for over 20 years, since he had first become involved with 'a valve machine'. He became the owner of a personal computer in 1978. Computing had added stability to his life at a time of crisis.

'I became genuinely addicted to computers in 1963 when I was personally miserable, it was a substitute. It's the answer to a lot of people's fantasies, you can dispense with people. My colleagues thought I was weird, but I missed out on things because I was an extreme neurotic, not because of computing. I would have been excessively reading, as I did before, or possibly worst becoming a social deviant of some sort.'

He still considered himself to be very shy and introverted, finding people judgemental, threatening and confusing. Relationships remained tense and difficult, although his wife had been able to offer him great stability and reassurance. He had met his wife when he was in his late twenties, but 'I was very immature at the time, she was not interested in boys', and it was a number of years before their relationship developed. Because of their jobs they, like some other professional couples, only live together at the weekends. This seemed to offer the ideal lifestyle to Frank, as he was able to immerse himself in his computing during the week and leave it at the weekend. There were therefore no conflicts within this marriage about the time he devoted to computing.

Case study G

'Explorer'
Male, early thirties, married, one daughter
Time spent on computer per week: 50+ at work

John saw his working-class background as happy on the whole, although there was far too much physical violence within the family for his liking;, between his parents and directed towards him. His mother was 'very much a mother, very loving and dependable, but she didn't encourage any of us (there were five children) to be independent. She needed to be a mother—needed to be adored'. His father, whom John respected a great deal, was 'loving, but not demonstrative'. He remembers being beaten by his father, after he had misbehaved and accused his parents of not loving him. This was his father's way of showing that he did.

John was not enthusiastic about his schooldays. 'I felt there were things which I had to do and didn't want to, and things I did, I didn't have a chance to.' He felt himself to have been very solitary at school. 'I only had two close friends (boys), I was terribly shy. I never had a girlfriend until I met Jane. I always felt very different, I was far too shy to have girlfriends. I didn't think they would be interested, although I got on well with girls. I felt I didn't know how to go about getting a relationship going. I was afraid, I think. Jane seduced me. I suppose I've always found it difficult to make friends. I wasn't the sort of person to go out and make friends. I hated parties and discos—I don't dance and people would stare at me.'

When asked to give an overall impression of himself John described himself as, 'Very shy. Lacking tremendously in confidence with people and my job. I feel I'm misunderstood and probably blame myself. I have an inability to express myself properly. I always felt very different—a loner. I wasn't interested in doing what my contemporaries were doing. I wondered what was wrong with me. I'm very much a romantic—I think I'm gentle, kind and caring. Some people see me as Jane's puppet—I don't appear to have the ability to make my own decisions.'

Having become bored with his first job, John enrolled on a government scheme as a trainee programmer, took a post with a computer company and rose to the level of analyst programmer. He loved the job. 'It was a challenge—solving problems—trying to beat deadlines, as opposed to meeting them. There was a fair degree of autonomy—you could write programs as you wanted as long as it did the job. On the whole I was happy with it—too happy with it, until it went sour.'

John differed from all of the other Dependent respondents in that he was also the only one who had never owned his own personal computer. He computed only at work or on friends' computers, and it appeared that much of his time at work was spent doing his own programming. 'Most was supposed to be for work, but I always tried to include exploration of the system in work. I spent far too long. Some of my programs were not as efficient as they should have been. I spent far too long mastering the system and playing games. My most ambitious project was to break down an adventure game—to crack it. It was a pirated, Honeywell game. I became totally hooked on games—the 'space-invader' type. I still find it difficult to walk past one. All my thoughts were totally turned over to computing, any spare time, even when working and every lunch time. I was a fanatic. I used to steal the housekeeping money to play games. I even used to drive 40 miles to a service station to play.'

But programming was his great love. 'I was always having to go back—I wondered if this was often deliberate—I never left it alone—it was never perfect—perfection meant it was finished. I used pen and paper at first, then I composed straight in—It's the way of getting as much reaction as possible. I tried to push a program to its limits, but wanted to find faults—the faults on the machine—I never questioned my programming abilities. It was absolutely wonderful and amazing. It's the thrill of mastery. If I'd had free rein I would have just explored it and not produced any work, but producing goods showed you could master it—being able to beat it, solve it, to prove your own powers over machines—to prove to myself that I was able to control one relationship.

I think I felt I needed something that I felt I was very good at. I felt I was a jack-of-all-trades—I felt I was *really* good at computing.'

John had frequently compared his interaction with the computer to that with people. 'People are so difficult, their actions are unpredictable. There is a risk in expressing yourself to others. I felt insecure with people—afraid of relationships being one way from my side—afraid they wouldn't like me. Computers are two-way and totally impersonal. I felt I could express my personality—felt I could interact as a person. It is totally predictable—its responses were totally dependent upon me and my actions. The feedback is great—it's very narcissic—when a program works it shows you are okay and alright. It ties in with my own feeling of insecurity with people.'

He used to spend at least two five-hour sessions per day on the computer at work, much of which was expected by his company. 'The work hours were officially 9 to 5, but they expected everyone to do extra for no money—the 'donkey and carrot' trick—there were no unions—if you work hard you'll get on—promises of great things to come. Jane and I shared a car. She often had to wait hours for me to finish work. I often worked on Saturday—she opposed all the extra work.'

Their marriage began to suffer badly, with both believing that computing was to blame for the conflicts and constant arguments which arose. John was torn between his work and Jane. 'She put a lot of pressure on me just to do my job and leave at five. There was always too much work, and Jane was very into Marx and exploitation. On one occasion Jane drove home and left me, it was miles from where we lived. Our social life revolved around computing—with other addicts—men-talk all day. I wanted—tried to explain to Jane. She didn't want to know, although she could have understood if she'd wished. I brought work home—pencil and paper stuff. I didn't have a micro but would have loved one. I used to gaze in shop windows—we were quite poor. Jane's influence stopped it—micros were more expensive in those days (the late 1970s).'

He stated that Jane had initially been very proud and encouraging of his computing abilities, especially by his rapid promotions and salary rises, but 'then she began to compare it with "another woman"—she changed from being supportive to absolute rage'.

'My happiest period was when I met Jane, and when I started working with computers. The worst time was when I realized I was hooked on computing and how it was affecting me—because I was living with Jane and because of the effect on the relationship, and me personally. I was just becoming less considerate of other people. I was just totally obsessed by it. Whatever I was doing I would rather have sat in front of a keyboard. Pubs were alright—I liked alcohol and smoking, and could play space invaders. When computing was strong, in my spare time I'd go to the loo! I was hooked and I knew it. Part of me knew I shouldn't be like that. I got adrenalin highs—hands shaking, knees wobbly. It's difficult to shut down the symptoms of excitement. I didn't think I liked the physical effects. I had a very high heart rate, but I had to go back. Games were easier to get an adrenalin fix. It's just a big trip all the time.'

Early in 1981 they separated, and John became ill. 'I wasn't sleeping—I was continuously hyped up, had the shakes, wasn't sleeping, wasn't working properly. I felt fear, apprehensive, fearful all the time. There was lots of conflict in my family at the same time—it was never just computing alone. I was always an anxious person before this.'

He had already been prescribed tranquillizers and anti-depressants, but a psychiatrist suggested hospitalization 'to reduce the very high adrenalin levels. I agreed to go in, as they promised I would not be drugged but given calm, caring treatment'. Although frightened he agreed to go. 'They immediately put me on a drip to cut down my adrenalin. I was drugged up to the eyeballs. I had no support from Jane as she had left. I had a bad reaction, my blood pressure dropped—I nearly died. I got really worried. After a week I wanted out, but was powerless. I felt I had to play it their way to even escape.' He remained in hospital for three weeks, and was absent from work for three months. Jane did return to him.

Their relationship improved dramatically as John learned to communicate his feelings, and through his wife's influence and interests he is now studying for a degree in Sociology. He says he is frightened of computers now. 'When I was ill I changed my computing ideas completely. I had to get out. I'm, a lot more laid back than I used to be, even with having the baby—Jane had an appalling labour—but I cope much better.'

Case study H

'Explorer'

Male, early forties, divorced, custody of two children

Hours spent computing per week: 72 hours at home

Bill had a very different educational background from the majority of the other respondents, and found school work difficult, preferring craft subjects and physical education. 'I was not academically inclined, I've never finished a book in my life.' He gained no formal qualifications and has had a very varied career, in and out of semi-skilled and unskilled work.

He did not feel that he mixed well at school. 'I was more shy than others, and find it hard to make conversation. I couldn't mix—I felt inferior to other people. Even now I have only one good friend.' Despite these comments he said that he did not find it difficult to share with those who did not make fun of him, but stated that he never had much of a social life and now had none at all, devoting his time to looking after his children.

His childhood he described as 'typically Jewish'. He saw his mother as very helpful and friendly—'I can tell her anything—we are so much alike we have arguments. She helps me with my problems and I knew she loved me. My father was cold and unloving. He never kissed or hugged me. I can't talk to him at all. He'd only help me now because of my mother. I don't know if he ever loved me.' When asked about his attitude towards his upbringing he stated: 'It was all wrong. There was not enough love given. They always wanted me to be clever, but I knew I wasn't.'

While still at school he started gambling seriously—an activity which lasted for 20 years. Even at school he lost a great deal of money. He gambled 'on anything—horses, dogs, at casinos . . . and regularly got into debt. I had no other interests, no interests in women. It's a sex substitute'. He viewed his gambling as 'nothing to do with money. It's the thrill, the excitement. I enjoyed losing more than winning. If I won £1000 there was nothing else to go for. If I lost it there was something to strive for, and the excitement.

'I got in such a state with my gambling I advertised for a woman to invest money in a business and get married . . . She lived with her grand mother. She gave me £2000 to get engaged, and £5000 plus to get married. I always wanted true love and someone to love—someone to put their arms around me—that's my ambition . . . I wan't happy in marriage. I love sex and my wife didn't—it was never happy. It was a kind of arranged marriage, we were never happy. I had nothing done for me—I had to do all the cleaning and cooking,. She didn't like sex—I didn't know her. I visited her once a week for six months before marriage. I knew on our honeymoon she was ill. She was like a child, not capable of looking after herself.' He felt that the only good things to come out of the marriage were 'the children and the money to get me out of trouble'. His wife spent much of their married years in and out of psychiatric hospitals with 'breakdowns', and he had to give up work to look after the children. After 10 years of marriage he eventually decided to divorce her, with the uncontested divorce and the custody of the children being settled very quickly.

Bill found it very difficult to describe himself. 'I'm moody. I couldn't care less what other people think . . . . That's a hard one . . . . I'm easy to get on with. I love my children—just my children. I'm an impulsive person, always thinking of ideas, inventions. I was happiest when I got divorced—I could do my own thing. It was a headache, I had to look after my wife and kids as well. When I lost a great deal of money I was unhappy . . . . Our second child died of a 'cot death'—it's the only time I ever cried.'

Computing entered his life in an unusual manner. 'Because I was so depressed about life, I wanted an interest. The social worker from the psychiatric hospital (his wife's) tried to help with ideas but I wasn't interested in any of them. I needed something to fill the evenings in particular—the days were busy, but I felt lonely. I prayed to God so hard for an interest—I was fighting against it (gambling)—I prayed so hard for an interest.' Shortly afterwards a friend showed him his microcomputer. 'I got straight into it. It amazed me. I thought I wouldn't be able to do anything. It took three years to get it (gambling) out of my system—the computer cured it. Without it I would be being treated for depression and I'd be a very unhappy person. I might have joined a *Gingerbread* group, but I'm not interested, I don't want to get involved.'

He bought his own computer in order 'to have a chess partner and for games—as an interest, a hobby—for teaching myself a new interest'. He spends most of his time 'programming, sorting out disks, teaching myself more programming and exploring the system. I started and never finished a big program, but I feel I could write anything I wanted to.' He did atually complete a games program but was unable to sell it, and still writes small programs for the children. His son enjoys using the computer but his daughter has shown little interest, but, as he added, 'they have benefited because I'm happy, most definitely'.

He estimates that he has about 4000 programs, only four of which he had bought; mainly games and 'chess, puzzles, quizzes, board games and adventure games, and lots of utilities, word processing packages and spreadsheets'. Of his collection he only uses a word-processing package and a few games. Despite living on social security he has spent about £1300 on his computing but felt 'no guilt attached. I'm careful with money, I don't drink, don't drive a car—I walk everywhere—and I don't go out'. All extra money was spent on computing and he did admit that his telephone bill was large because of networking. 'It's a bit like gambling. I want more money to buy more (computer) things.'

His programming style was hands-on and unplanned. 'I've never written flow charts. I've spent weeks on one program—perfection is very important.' He greatly enjoys debugging his programs—'I had one program not quite working. I thought about it in bed—I had to get up. It's great when it works. It's the thrill of finding out the versatility of the computer, there's always something new to find out.'

'I'm hooked because it's different. It's a challenge. Every time there is something new to explore. Even on one particular adventure (game) there is always a different way of doing things. It's addictive to me—because I'm the type of person who likes a chance to do something different and exciting—where there are puzzles. I like the ability to create something on my own—something I can show someone else, perhaps even sell my ideas. Every time there is another challenge—there is always something different. It's my life. It's something I think about during the day—think about doing at night—it might be like gambling. It's a friend, a colleague—as unusual as it may seem.'

He found it difficult to limit the time he spent on the computer, but did not compute while the children were at home awake. 'I have to know all the chores are done, with everyone happy and relaxed.'

Computing had brought much happiness and enjoyment into his life. Recently he was accepted on a part-time, two-year course to teach BASIC programming in adult evening centres, and saw this as an ideal job. He would have loved to have been a primary school teacher, and he was already spending two afternoons a week giving voluntary help at his childrens' school—'I'm known as the Computer Man.' He had also made new friends over the networks, and people contacted him for advice.

'I feel relaxed—I don't know what to do first. It gives me satisfaction and a certain amount of excitement. I see myself becoming a teacher and teaching computing to primary schools. If I fail I just want to grow old gracefully. I feel I've lived a hundred years and I'm only forty.'

Appendix 2
Results from interviews with two sculptors

Interviews were carried out with two sculptors in order to examine whether and how their dedication to their work and art differed from the Dependents' dedication to computing. The results showed that although the actual hobbies and interests of the sculptors and the Dependents differed, there were remarkable similarities between them in other areas. The sculptors and the Dependents exhibited very similar attitudes and dedication to their work and play, as well as similarity in their personalities, social experiences, and family backgrounds. The following notes summarize the results from the interviews with the sculptors.

Demographic data

(1) Both male; one in his late forties, the other in his early fifties.
(2) Both were divorced men with two children; one had since remarried.

Early hobbies

(1) *No interest in science, mechanics, electronics or computing* (the major difference between the Dependents and the sculptors).
(2) Little interest in team sports or formal social activities.
(3) Had always been extremely interested in their chosen hobbies from an early age.
(4) Most of their hobbies had been connected with art in some form, with both spending much of their childhood drawing and sketching.
(5) Their hobbies were invariaby undertaken in solitude.

Work/play

(1) They did not easily differentiate between work and play.
(2) Their main leisure interest had become a means of earning a living.

(3) They spent a considerable amount of time sculpting (although not usually as much as the Dependents spent computing—physical fatigue hampered them more than the lack of will to continue).
(4) Elements of fantasy were important to their work and were able to be expressed.
(5) They needed to gain control and have power over their media; feelings of 'playing God' with their fantasies and creations.
(6) They saw themselves as pioneers.
(7) They worked hands-on, often with little pre-planning, allowing the medium with which they were working to direct the development of their ideas and creations.
(8) The *process* of sculpting was more important than the end-product.
(9) They lost track of time whilst sculpting, and when focused on their activities they were insensitive to the world around them.
(10) They frequently did not know when to stop—some pieces were never good enough, never perfect; more could always be done.
(11) They could lose interest in a piece when their ideas led them to try something new.
(12) They possessed a great deal of unsold and often unfinished work to which they felt they might return.
(13) Interest was intrinsic to the work, not extrinsic to the sale of the object (the process of selling their work was a necessary chore not a pleasure).
(14) They spent a great deal of money on their activities, buying good tools and expensive materials (high quality woods and metals, rather than computer equipment).
(15) Most of their money was tied up in their sculpting; other more practical items were often ignored.

Work and other people

(1) They needed to work alone, but enjoyed helping and advising others.
(2) They often found it difficult to describe their work and their enjoyment to the layman; they only had a few like-minded friends with whom they could really share their ideas.
(3) They gained status and prestige from their abilities with some laymen, although they felt they were considered eccentric by others.

Social and psychological

(1) They considered themselves to have been 'loners' and shy from an early age, and felt isolated from their peers while at school.
(2) They considered themselves to have poor social and verbal skills.

(3) They did not understand what others required from relationships.
(4) They found many social encounters difficult.
(5) They were uninterested in the non-focused conversations of people-oriented persons, preferring conversation to be task-centred.
(6) Their spouses were said to have felt alienated from them and their work, with one stating that his wife accused him of being wedded to his other woman, 'art'.
(7) They found relationships easier with women than men and needed female company.
(8) They were a little bewildered about the meaning of life, and sometimes depressed.
(9) They found materials easier to work with than people.
(10) Their parents were considered to have been distant and not overtly affectionate (one father was described as very authoritarian and strict).
(11) As children they were rewarded for their achievements; they felt that their parents' love had been conditional and withdrawn at times.
(12) They had always been object-centred throughout life; they were able to prove their worth by the achievement of tasks.
(13) They gained their greatest satisfaction from their sculpting, it removed them from the frustrations and uncertainties of life and gave them peace and security.

As so many similarities were observed between the sculptors and the Dependents, the sculptors were asked to complete the same tests as had been given to the interviewed Dependents (the GEFT and the 16PF). Because only two sculptors took part in this small investigation, no comparative statistical analyses were carried out between the sets of results and the discussion relies upon gross observation of the data only.

Results from the GEFT

Both sculptors were found to be very field-independent as measured by the GEFT. Neither made any mistakes, completing the 18 items correctly, and each completed the two sections of the test well within the time limit of 10 minutes (at 5 min 35 s and 7 min 10 s respectively). The results were therefore very similar to those of the Dependents.

Such results are consistent with the self-descriptions given by the sculptors, as high levels of field-independence have been shown to indicate:

- An ability to deal with symbolic representations (essential for their type of art which was symbolic, often fanciful, rather than realistic);
- A well developed sense of identity for those little influenced by the views of others;
- The use of isolation as the coping mechanism;
- A cognitive style thought to develop from parenting which encourages separate autonomous functioning.

Results from the 16PF

Comparisons made between the 16PF profiles of the sculptors and the Dependents showed that there were many similarities between them (see Figure A2.1) Both groups yielded equivalent scores for the First Order factors A, B, C, F, G, L, M, O, Q1, Q2, Q3, and Q4, and for the Second Order factors QII and QIV. These results suggested that both groups were as equally reserved, intelligent, imaginative, self-sufficient and independent as each other. Like the Dependents, the sculptors' results indicated that they also were the type of people who preferred to have one or two close friends, did not enjoy large gatherings of people, and were used to making their own decisions without recourse to others and public opinion (Cattell *et al.*, 1970).

Some differences for the First Order factors were observed however, and the sculptors were shown to be less assertive (factor E), more shy (factor H), more forthright (factor N), and far more tender-minded (factor I) than the Dependents. The sculptors' high scores for factor I suggest that they were more emotionally sensitive, intuitive, artistic, temperamental and more likely to be day-dreamers and less realistic than the Dependents. Their lower scores on factor N suggested that they were more 'forthright, natural, genuine, unpretentious' than the Dependents, with a 'natural warmth and a genuine liking for people' (IPAT, 1979).

The differences between the sculptors and the Dependents within the First Order factors led to Second Order differences also. Because of the manner in which factor QI introversion : extroversion is determined, as the sculptors were less assertive and more shy than the Dependents they were also shown to be more introverted than the Dependents. Similarly, the sculptors were found to exhibit less tough poise (factor QIII) than the Dependents because of their very high scores for factor I and low scores for factor N.

Discussion

The sculptors' 16PF results were in keeping with their more inward-looking, spiritually-private, less intellectualized lives, but showed that they also bore great similarity to the Dependents. Their activities did not deal with technologically-manufactured artefacts, but with the manipulation and command of raw materials which they shaped into images which had no functional purpose other than to be aesthetically pleasing or intellectually challenging to the eye and emotions. As with the Dependents, the process of undertaking their work was of equal if not more importance than the finished product from their labours, and the need for control over their media was just as important to the sculptors as to the Dependents and engendered the same feelings of dominance.

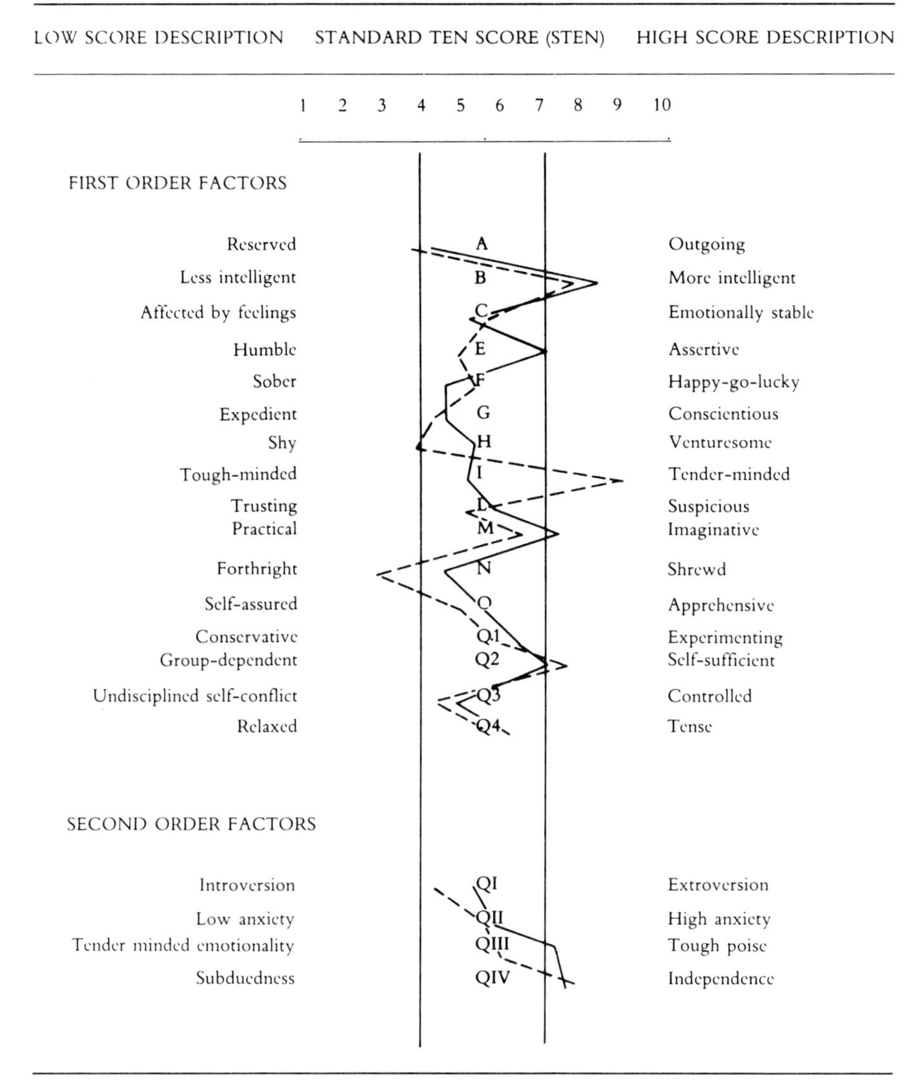

Figure A2.1 Mean 16PF sten scores of the Dependents and the sculptors (vertical lines define 1 S.D. from population norm).
Dependents ______ (N = 45), Sculptors _ _ _ _ (N = 2)

'Artists' are by tradition not expected to be field-independent, nor to need to control their environment by discovering the logical relationships between things in an analytical, deductive manner, but are expected to be divergent, imaginative, creative thinkers. The two sculptors, both of whom were successful and professional artists, exhibited both the facets of field-independence and creativity in their personalities, and utilized them within their work. From this, one would assume in order to work successfully in three dimensions with a variety of media and techniques that field-independence would be a significant asset, and would not prove to be incompatible with imaginative creativity.

The sculptors considered themselves to have been object-centred from an early age, and like the more scientific Dependents were not averse to using their preferred activities as an escape from problems within relationships. Why the sculptors had become 'artists' and not 'scientists' is unclear, although Roe (1957) discovered there to be no correlation between object-centredness and an aptitude for science. Object-centredness can apparently take very different forms. The results from the two sculptors suggested that their object-centredness had developed at an early age, leading to an intense interest in one particular area of activity. This could have been attributed to parenting methods, which in addition could have created a certain amount of distrust in humans and the high level of self-sufficiency and independence as exhibited by the Dependents.

Having assumed that I had selected two people with distinctly different attitudes and needs from the Dependents, the results showed great agreement between them, suggesting that there may be more similarities between the scientist and the artist than one might initially suppose. The similarities between the sculptors and the Dependents may of course be purely coincidental. Further research would be needed in order to establish whether the object-centred personality can be firmly attributed to early child-rearing habits; and whether it invariably leads to the development of an intense interest in one particular hobby to the exclusion of an eclectic range, whether of an artistic or scientific nature. These interviews, together with many discussions with engineers and computer scientists, however, did seem to suggest that there is more than an element of truth to the generalisability of these suppositions.

Bibliography

Allport, G. W. (1963). *Pattern and growth in personality*. (London: Holt, Rinehart and Winston).

Altman, I. and Taylor, D. A. (1973). *Social penetration: the development of interpersonal relationships*. (London: Holt, Rinehart and Winston).

Altus, W. D. (1965). Birth order and primogeniture. *Journal of Personality and Social Psychology*, **2**, 872–876.

American Psychiatric Association. (1980). *Diagnostic and statistical manual of mental disorders*, 3rd edition. (Washington DC: American Psychiatric Association).

APEX. (1985). *New technology: a health and safety report*. (London: APEX).

ASTMS. (1979). *Guide to health hazards of visual display units*. ASTMS Policy Document, London.

Atkinson, J. W. and Raynor, J. O. (1974). *Motivation and Achievement*. (London: John Wiley).

Auden, W. H. (1948). Squares and oblongs. In *Play: its role in development and evolution*, edited by J. S. Bruner, A. Jolly and K. Sylva (Harmondsworth: Penguin) (1976).

Baker, J. R. (1939). Counterblast to Bernalism. *New Statesman*, 29th July.

Barclay, A. G. and Cusumano, D. (1976). Father absence, cross-sex identity, and field independent behaviour in the male adolescents. *Child Development*, **38**, 243–250.

Begole, C. (1984). The women's market emerges. *Software merchandising*, June, 54–61.

Bell, E. G. (1955). Inner-directed and other-directed attitudes, PhD thesis, Yale University. In Witkin, H. A., Oltman, P. K., Raskin, E. and Karp, S. A. (1971). *A manual for the Embedded Figures Test*. (Palo Alto, CA: Consulting Psychologists Press Inc.)

Beloff, H. (1985). *Camera culture*. (Oxford: Basil Blackwell).

Bene, E. (1965). On the genesis of male homosexuality. *British Journal of Psychiatry*, **111**, 815–821.

Berdie, R. F. (1944). Factors related to vocational interests. *Psychological Bulletin*, **41**, 137–157.

Bernal, J. D. (1939). Professor Bernal replies. *New Statesman*, 5th August.

Berqvist, U. (1984). Video display terminals and health. *Scandinavian Journal of Work, Environment and Health*, **10**, supplement 2.

Bieber, I. (1962). *Homosexuality: a psychoanalytic study*. (New York: Basic Books).
Biller, H. B. (1969). Father absence. Maternal encouragement and sex role development in kindergarten-age boys. *Child Development*, **40**, 539–546.
Biller, H. B. (1971). *Father, child and the sex role*. (Lexington, MA: Health Books).
Biller, H. B. (1971). The mother-child relationship and the father absent boy's personality. *Merrill-Palmer Quarterly*, **17**, 227–241.
Bishoff, D. (1983). *War Games*. (Harmondsworth: Penguin).
Bishop, D. W. and Witt, P. A. (1970). Sources of behavioural variance during leisure time. *Journal of Personality and Social Psychology*, **16**, 352–360.
Blake, W. (1979). *The book of Urizen*. (London: Thames and Hudson).
Blanchard, R. W. and Biller, H. B. (1971). Father availability and performance among third grade boys. *Developmental Psychology*, **4**, 301–305.
Blendis, J. (1982). Men's experiences of their own fathers. In *Fathers: psychological perspectives*, edited by N. Beail and J. McGuire, (London: Junction Books).
Boden, M. A. (1977). *Artificial intelligence and natural man*. (Brighton: Harvester Press).
Boden, M. A. (1981). *Minds and mechanisms*. (Brighton: Harvester Press).
Boden, M. A. (1981). The meeting of man and the machine. In *The design of information systems for human beings*, edited by K. P. Jones and H. Taylor (London: ASLIB).
Borrill, J. and Reid, B. (1986). Are British psychologists interested in sex differences? *Bulletin of the British Psychological Society*, **39**, 286–288.
Bowden, Lord, B. V. (1965). *New Scientist*, 30th September, 849.
Bowlby, J. (1973). *Attachment and loss*. II Separation, anxiety and anger. (London: The Hogarth Press).
Bowman, R. F. (1982). The 'Pac-Man' theory of motivation: tactical implication for classroom instruction. *Educational technology*, **22**, 9, 14–16.
Brabant, S. (1982). The computer as a construct of social reality. *SIGUGGS Newsletter (USA)*, **12**, 3, 12–17.
Brockner, J. (1979). The effects of self esteem, success-failure, and self-consciousness on task performance. *Journal of Personality and Social Psychology*, **37**, 1732–1741.
Brod, C. (1984). *Technostress: the human cost of the computer revolution*. (Reading, MA: Addison-Wesley).
Bronfenbrenner, U. (1961). Some familial antecedents of responsibility and leadership in adolescents. In *Leadership and interpersonal behaviour*, edited by L. Petrullo and B. M. Bass (New York: Holt, Rinehart and Winston).
Bruner, J. S., Jolly, A. and Sylva, K. (1976). *Play: its role in development and evolution*. (Harmondsworth: Penguin).
Burchfield, R. W. (ed). (1972). *Supplement to the Oxford English Dictinary*. Vol. 1. (Oxford: Oxford University Press).
Burchfield, R. W. (ed). (1976). *Supplement to the Oxford English Dictionary*. Vol. 2. (Oxford: Oxford University Press).
Burchfield, R. W. (ed). (1982). *Supplement to the Oxford English Dictionary*. Vol. 3. (Oxford: Oxford University Press).
Buros, O. K. (1972). *The seventh mental measurement year book*. Vol. 2. (New Jersey: Gryphon Press).
Cakir, A., Hart, D. J. and Stewart, T. F. M. (1980). *Visual display terminals*. (Chichester: John Wiley).

Capra, P. C. and Dites, J. E. (1962). Birth order as a selective factor among volunteer subjects. *Journal of Abnormal and Social Psychology*, **64**, 302.

Card, W. I., Nicholson, M., Crean, G. P., Watkinson, G., Evans, C. R., Wilson, J. and Russell, D. (1974). A comparison of doctor and computer interrogation of patients. *International Journal of Biomedical Computing*, **5**, 175–187.

Carlisle, J. H. (1974). *Man computer interaction problem solving relationships between user characteristics and interface complexity*. PhD thesis, Yale University. (Ann Arbor, MI: University Microfilms), No. 74–25725.

Carlyle, T. (1829). Signs of the times. In R. Williams (1967). *Culture and society 1780–1950*. (London: Chatto and Windus).

Carroll, J. M. (1982). The adventure of getting to know a computer. *Your Computer*, 5, 11, 47–58.

Catt, I. (1971). *Computer worship*. (London: Pitman).

Cattell, R. B., Eber, H. W. and Tatsuoka, M. M. (1970). *Handbook for the Sixteen Personality Factor Questionnaire (16 PF)*. (Champaign, IL: Institute for Personality and Ability Testing).

Check, J. M., Perlman, D. and Malamuth, N. (1985). Loneliness and aggressive behaviour. *Journal of Social and Personal Relationships*, **2**, 243–252.

Chick, J., Waterhouse, L. and Wolff, S. (1979). Psychological construing in schizoid children grown up. *British Journal of Psychiatry*, **135**, 425–430.

Clarricoates, K. (1978). Dinosaurs in the classroom. *Womens Studies International Quarterly*, **1**, 353–364.

Cockburn, C. (1985). *Machinery of dominance: women, men and technical know-how*. (London: Pluto Press).

Cockcroft, J. D. (1985). *An investigation into the difference in responses to a postal type questionnaire*. Unpublished BSc project report, Loughborough University.

Committee of enquiry into the teaching of mathematics in schools. (1982). *Mathematics Count*. (London: HMSO).

Committee on the Future of Broadcasting, N. Annan, Baron (Chairman). (1977). *Report of the future of broadcasting*. Cmnd. 6753. (London: HMSO).

Cornwall, H. (1986). *The new hacker's handbook*. (London: Century).

Cox, F. N. (1962). An assessment of children's attitudes towards parental figures. *Child Development*, **33**, 821–830.

Craig, C. (1986). Wired for equality. *The Guardian*, 11th December, 16.

Cull, A., Chick, J. and Wolff, S. (1984). A consensual validation of schizoid personality in childhood and adult life. *British Journal of Psychiatry*, **144**, 646–648.

Cullingford, C. (1984). *Children and television*. (Aldershot: Gower).

Culpin, M. (1948). *Mental abnormality*. (London: Hutchinson University Library).

Department of Education and Science. (1985). *Statistics of education: school leavers CSE and GCE 1985*. (London: HMSO).

Dion, K. K. and Berscheid, E. (1974). Physical attractiveness and peer perception among children. *Sociometry*, **37**, 1–12.

Drever, J. (1952). *A dictionary of psychology*. (Harmondsworth: Penguin).

Duncan, A. M. (1986). *The splitting of the culture*. Unpublished, Loughborough University.

Dweck, C. S., Davidson, W., Nelson, S. and Enna, B. (1978). Sex differences in learned helplessness: II. The contingencies of evaluative feedback in the classroom, and III. An experimental analysis. *Developmental Psychology*, **14**, 268–276.

Eason, K. D. (1976). Understanding the naive computer user. *The Computer Journal*, **19**, 3–7.

Edinger, J. A. and Patterson, M. L. (1983). Nonverbal involvement and social control. *Psychological Bulletin*, **93**, 30–56.

Egli, E. A. and Meyers, L. S. (1984). The role of video games in adolescent life: is there reason to be concerned? *Bulletin of the Psychonomic Society*, **22**, 309–312.

Eisele, J. E. (1981). Computers in the schools: now that we have them...? *Educational Technology*, **21**, 10, 24–27.

Elliot, J. (1974). Sex role constraints on freedom of discussion: a neglected reality of the classroom. *New Era*, **55**, 147–155.

Elliott, R. (1961). Interrelationships among measures of field dependence, ability, and personality traits. *Journal of Abnormal and Social Psychology*, **63**, 27–36.

Equal Opportunities Commission. (1983). *Information technology in schools. Guidelines of good practice for teachers of I.T.* (London: Borough of Croydon).

Equal Opportunities Commission. (1984). *WISE 1984—Women into science and engineering*. 371–425. (London: HMSO).

Equal Opportunities Commission. (1985). *Infotech and gender: an overview*. (Manchester: Equal Opportunities Commission).

Eysenck, H. J. (1977). *You and neurosis*. (London: Temple-Smith).

Favaro, P. J. (1982). Games for cooperation and growth: an alternative for designers. *Softside (USA)*, 6, 2, 18–21.

Fennema, E. (1981). Attribution theory and achievement in mathematics. In *The development of reflection*, edited by S. R. Yussen, (New York: Academic Press).

Ferguson, D. A. (1987). 'RSI': Putting the epidemic to rest. *Medical Journal of Australia*, 117, 213–214.

Firschein, O., Fischler, M. A., Coles, L. S. and Tenenbaum, J. M. (1973). Forecasting and assessing the impact of artificial intelligence on society. In *3rd International Conference on Artificial Intelligence*, 105–120.

Fitter, M. J. (1978). *Towards more 'natural' interactive systems*. MRC Memo 253, Sheffield.

Forer, L. (1977). *The birth order factor*. (New York: Pocket Books).

Forrest, P. (1984). Hello high school, bye-bye micro. *Primary teaching and micros*, September.

Forster, M. C. (1933). *A study of father-son resemblance in vocational interest patterns*. Unpublished MA thesis, University of Minnesota, cited by Nias, D. K. B. (1975).

Freud, S. (1947). Leonardo da Vinci: a study in pyschosexuality. (New York: Random House), cited by Carroll, M. P. (1978). *Behavioral Science Research*, **13**, 255–271.

Fromm, E. (1941). *Escape from freedom*. (New York: Farrar and Rinehart).

Frude, N. (1983). Hacking with the hackers. *New Scientist*, 24th February, 532–533.

Frude, N. (1983). *The intimate machine: close encounters with new computers*. (London: Century Publishing).

Furnham, A. (1981). Personality and activity preference. *British Journal of Social*

Psychology, **20**, 57–68.
Furnham, A. (1981). Personality and the choice and avoidance of social situations. In *Social situations*, edited by A. Furnham and A. Graham, (Cambridge: Cambridge University Press).
Galton, F. (1874). *English men of science: their nature and nurture*. (London: MacMillan).
Getzels, J. W. and Jackson, P. W. (1962). *Creativity and intelligence*. (New York: John Wiley).
Gibb, G. D., Bailey, J. R., Lambrith, T. T. and Wilson, W. P. (1983). Personality differences between high and low electronic video game users. *Journal of Psychology*, **114**, 159–165.
Gjerde, C. M. (1949). *Parent-child resemblances in vocational interests and personality trails*. Unpublished PhD thesis, University of Minnesota, cited by Nias, D. K. B. (1975).
Grandjean, E. (Ed.) (1984). *Ergonomics and health in modern offices*. (London: Taylor and Francis).
Grandjean, E. and Vigliani, E. (Eds.) (1980). *Ergonomic aspects of visual display terminals*. (London: Taylor and Francis).
Gray, K. (1984). *Implication of microcomputers for counsellors*. (Guildford: Guildford County College of Technology).
Gray, S. (1959). Perceived similarity to parents and adjustment. *Child Development*, **30**, 91–107.
Greenfield, P. M. (1984). *Mind and media*. (London: Fontana).
Guildford, J. P., Christensen, P. R., Bond, N. A. and Sutton, M. A. (1954). A factor analysis study of human interests. *Psychological Monographs*, **68**, 375.
Guntrip, H. (1968). *Schizoid Phenomena, object relations and the self*. (London: Hogarth Press).
Hacker Papers, The. (1980). *Psychology Today*, August, 62–70.
Haddon, L. (1986). *Some gender issues around micros*. Unpublished working paper Imperial College, London.
Hagedorn, R. and Labovitz, S. (1968). Participation in community associations by occupations. *American Social Review*, **33**, 272–283.
Hall, E. G. and Lee, A. M. (1981). Sex and birth order in children's goal setting. *Perception and Motor Skills*, **53**, 663–666.
Halloran, J. D. (1964). Television and violence. *Twentieth Century*, Winter, 61–77.
Hammond, W. H. (1945). An analysis of youth centre interests. *British Journal of Educational Psychology*, **15**, 122–126.
Harris, P. R. (1984). Shyness and psychological imperialism. *European Journal of Social Psychology*, **14**, 169–181.
Haun, P. (1965). *Recreation: a medical viewpoint*. (New York: Columbia University).
Haynes, J. R. and Carley, J. W. (1970). Relation of special abilities and selected personality traits. *Psychological Reports*, **26**, 214.
Hebb, D. O. (1955). Drives and the CNS (conceptual nervous system). *Psychological Review*, **62**, 243–254.
Hertzberg, F. (1966). *Work and the nature of man*. (London: Staples Press).
Hesse, H. (1973). *Siddhartha*. (London: Pan Books).
Himmelweit, H. T., Oppenheim, A. N. and Vince, P. (1958). *Television and the child*. (Oxford: Oxford University Press).
Hite, S. (1988). *Women and love*. (London: Viking).

Hoffman, L. W. (1961). The father's role in the family and the child's peer-group adjustment. *Merrill-Palmer Quarterly*, **7**, 97–105.

Hofstadter, D. R. (1979). *Godel, Eschel, Bach: an eternal golden braid.* (Brighton: Harvester Press).

Hollands, J. (1985). *Silicon Syndrome: How to survive a high-tech relationship.* (New York: Bantam Books).

Horner, M. S. (1970). Femininity and successful achievement: a basic unconsistency. In *Feminine personality and conflict*, edited by J. M. Bardwick, (Monterey, CA: Brooks/Cole).

Horney, K. (1946). *Our inner conflicts: a constructive theory of neurosis.* (London: Routledge and Kegan Paul).

Horney, K. (1951). *Neurosis and human growth: the struggle toward self-realization.* (London: Routledge and Kegan Paul).

Howitt, D. and Cumberbatch, G. (1975). *Mass media, violence and society.* (London: Elek Science).

Hoyle, F. (1968). *Physics Today*, **21**, 4, 148.

Hudson, L. (1966). *Contrary imaginations: a psychological study of the English schoolboy.* (London: Methuen).

Huizinga, J. (1976). Play and contest as civilizing functions. In *Play: its role in development and evolution*, edited by J. S. Bruner, A. Jolly and K. Sylva, (Harmondsworth: Penguin).

Hunt, J. M. (1965). Intrinsic motivation and its role in psychological development. In *Nebraska Symposium on Motivation Vol. 13*, edited by D. Levine, (Nebraska: Nebraska University Press).

Hutt, C. and Bhavnani, R. (1976). Predictions from play. In *Play: its role in development and evolution*, edited by J. S. Bruner, A. Jolly and K. Sylva, (Harmondsworth: Penguin).

IPAT. (1979). *Administrator's Manual for the 16PF.* (Champaign, IL: Institute for Personality and Ability Testing).

Isherwood, J. (1987). Psychosis warning to computer freaks. *The Daily Telegraph*, 2nd September, 5.

Jacobson, G. and Ryder, R. G. (1969). Parental loss and some characteristics of the early marriage relationship. *American Journal of Orthopsychiatry*, **39**, 779–787.

Jones, W. H., Freeman, J. E. and Goswick, R. A. (1981). The persistence of loneliness: self and other determinants. *Journal of Personality*, **49**, 27–48.

Jourard, S. M. (1971). *Self disclosure: an experimental analysis of the transparent self.* (London: John Wiley).

Kaplan, D. M. (1972). On shyness. *International Journal of Psychoanalysis*, **53**, 439–453.

Kelly, A. (1982). *The missing half.* (Manchester: Manchester University Press).

Kelly, A. (1982). Why girls don't do science. *New Scientist*, 20th May, 497–500.

Kerr, A. (1987). Care required as introverts take to industry. *The Sunday Times*, 10th May, 85.

Keys, W. and Ormerod, M. B. (1976). A comparison of the pattern of science subject choices for boys and girls in the light of pupils' own expressed subject preferences. *School Science Review*, **58**, 343–350.

Kidder, T. (1981). *The soul of a new machine.* (London: Allen Lane).

Kiesler, S. and Sproull, L. (1984). *Response effects in the electronic survey.* (Pittsburgh,

PA: Carnegie-Mellon University).
Kiesler, S., Siegal, J. and McGuire, T. W. (1984). Social psychological aspects of computer-mediated communication. *American Psychologist*, **39**, 1123–1134.
Kiesler, S., Seigel, J. and McGuire, T. W. (1987). Social psychological aspects. In *Information technology: social issues*, edited by R. Freeman, G. Salaman and K. Thompson, (Sevenoaks: Hodder and Stoughton).
Kirkland, J. (1976). Interest: Phoenix in psychology. *Bulletin of the British Psychological Society*, **29**, 33–41.
Klemmer, E. T. and Snyder, F. W. (1972). The measurement of time spent communicating. *The Journal of Communication*, **22**, 142–158.
Kotelchuck, M., Zelano, P., Kagan, J. and Spelke, E. (1975). Infant reaction to parental separation. *Journal of Genetic Psychology*, *126*, 255–262.
Krug, S. E. (1981). *Interpreting 16PF profile patterns.* (Champaign, IL: Institute for Personality and Ability Testing).
Lake, F. (1966). *Clinical theology*. (London: Darton, Longman and Todd).
Large, P. (1984). *The micro revolution revisited.* (London: Frances Pinter).
Laurie, P. (1980). *The micro revolution: A change for the better or for the worse?* (New York: Futura Publications).
Leary, M. R. and Dobbins, S. E. (1983). Social anxiety, sexual behaviour and contraceptive use. *Journal of Personality and Social Psychology*, **47**, 775–794.
Leiderman, G. F. (1959). Effects of parental relationships and child-training practices on boys' interactions with peers. *Acta Psychologica*, **15**, 469.
Levy, S. (1984). *Hackers: heroes of the computer revolution.* (New York: Anchor Press/Doubleday).
Litwinski, L. (1950). Constitutional shyness. *Journal of General Psychology*, **42**, 299–311.
Loftus, G. R. and Loftus, E. F. (1983). *Mind at play: the psychology of video games*, (New York: Basic Books).
Machin, D. (1984). Mud, Mud, glorious Mud. *TeleLink*, May, 46.
MacKay, D. (1975). *Clinical psychology: theory and therapy*. (London: Methuen).
McClelland, D. C., (1962). On the psychodynamics of creative physical scientists. In *Contemporary approaches to creative thinking*, edited by H. E. Gruller, G. Terrell and M. Wertheimer, (New York: Atherton Press).
McClelland, D. C., Atkinson, J. W., Clark, R. A. and Lowell, E. L. (1953). *The achievement motive*. (New York: Appleton-Century-Crofts).
McClure, R. F. and Mears, F. G. (1984). Video games players: personality characteristics and demographic variables. *Psychological Reports*, **55**, 271–276.
McCorduck, P. (1979). *Machines who think*. (San Francisco: W. H. Freeman).
McIntyre, J. J. and Teevan, J. J. (1972). Television violence and deviant behaviour. In *Television and adolescent aggression*, Vol. 3, edited by G. A. Comstock and E. A. Rubenstein, (Washington, DC: US Government Printing Office).
McKechnie, G. E. (1974). The psychological structure of leisure; past behaviour. *Journal of Leisure Research*, **6**, 27–45.
Mahoney, P. (1985). *Schools for the boys?* (London: Hutchinson).
Malatesta, C. Z. (1985). Developmental course of emotion expression in the human infant. In *The development of expressive behaviour*, edited by G. Zivin, (New York: Academic Press).
Malde, B. (1978). *Further research in double interaction: the simultaneous conduct of*

man-made and man-computer interaction. Unpublished, PhD thesis, Loughborough University.

Malmquist, C. P. (1970). Depression and object loss in acute psychiatric admissions. *American Journal of Psychiatry*, **126**, 1782–1787.

Malone, T. W. (1984). What makes computer games fun? *Byte*, **6**, 258–277.

Martin, J. and Norman, A. R. D. (1970). *The computerized society.* (London: Penguin).

Maslow, A. H. (1970). *Motivation and personality.* 2nd edition. (New York: Harper).

Meddis, R. (1980). Unified analysis of variance by ranks. *British Journal of Mathematical and Statistical Psychology*, **33**, 84–98.

Meddis, R. (1984). *Statistics using ranks: A unified approach.* (Oxford: Basil Blackwell).

Meek, B. (1972). Computers and education. In *Computers and the year 2000*, edited by Lord Avebury, R. Coverson, J. Humphries and B. Meek, (Manchester: NCC Publications).

Mendelson, L. (1983). U.S. prepare to combat the electronic sweatshop image. *Computing*, 21st July, 17.

Miller, J. (1988). ITV's shows turn off the audience. *The Sunday Times*, 16th May, C6.

Mintell Leisure Intelligence. (1984). *Home Computers.* Issue 6. (London: Mintell).

Mitchell, D. and Wilson, W. (1967). Relationship of father absence to masculinity and popularity of delinquent boys. *Psychological Reports*, **20**, 1173–1174.

Mitchell, J. V. (1983). *Tests in print III.* (Nebraska: The Buros Institute of Mental Measurement, University of Nebraska Press).

NALGO Health and Safety Briefing. (1986). (London: NALGO).

Nash, M., Allsop, T. and Woolnough, B. (1985). *Factors affecting the uptake of technology in schools.* (Oxford: Department of Educational Studies).

Neubauer, P. (1960). The one-parent child in his oedipal development. *Psychoanalytic Study of the Child*, **15**, 286–309.

Neulinger, J. and Raps, C. S. (1972). Leisure attitudes of an intellectual elite. *Journal of Leisure Research*, **4**, 196–207.

Nias, D. K. B. (1975). *Personality and other factors determining the recreational interests of children and adults.* Unpublished PhD thesis, University of London.

Nicholson, J. (1984). *Video games-threat or challenge? A preliminary report.* Unpublished, University of London.

Norwood, R. (1986). *Women who love too much.* (London: Arrow).

O'Halloran, A. (1954). *An investigation of personality factors associated with underachievement in arithmetic and reading.* Unpublished MSc thesis, Purdue University. Cited by Cattell *et al.* (1970), *Handbook for the 16PF Questionnaire, (Champaign, IL: IPAT).*

Oberg, K. (1966). Cultural shock: adjustment to new cultural environments. *Practical Anthropology*, **7**, 177–182.

Oltman, P. K., Raskin, E., Witkin, H. A. and Karp, S. A. (1971). *GEFT: adaptation of the individually administered Embedded Figures Test.* (Palo Alto, CA: Consulting Psychologists Press Inc).

Ong, C. N., Koh, D. and Phoon, W. O. (1988). Review and reappraisal of health hazards of display terminals. *Display* 9, 1.

Orford, J. (1985). *Excessive appetites: a psychological view of addictions*, (Chichester: John Wiley).

Ormerod, M. B. and Billing, K. (1982). A six orthogonal factor model of adolescent personality. *Personality and Individual Differences,* **3,** 2, 107–117.

Osgood, C. E., Suci, G. J. and Tannenbaum, P. H. (1957). *The measurement of meaning*. (Urbana, IL: Illinois University Press).

Papert, S. (1972). A computer laboratory for elementary schools. *Computers and Automation*, **21**, 6, 19–23.

Papert, S. (1972). Teaching children thinking. *Program Learning and Educational Technology*, **9**, 245–255.

Papert, S. (1980). *Mindstorms: children, computers and powerful ideas*. (Brighton: Harvester Press).

Parliamentary Debates (Hansard). (1981). *Control of space invaders and other electronic games*. Sixth Series Vol. 5. May 18–June 5. (London: HMSO).

Pateman, T. (1981). Communicating with computer programs. *Language and Communication*, **1**, 3–12.

Pearce, B. G. (1983). *Ergonomics and the terminal junkie*. (Loughborough: HUSAT Research Centre).

Pearce, B. G. (1983). True ergonomic design for users: more than just furniture. *Ergonomics*, **25**, 5, 6–8.

Pearce, B. G. (1984). Trade unions and ergonomic problems. In *Health hazards of VDTs?* edited by B. G. Pearce, (Chichester: John Wiley).

Pearce, B. G. (1984). Where does ergonomics fit? *Management Review,* supplement to *Computer Weekly*, 19th January.

Pile, S. (1980). *The book of heroic failures*. (London: Futura).

Pilkonis, P. A. (1977). The behavioural consequences of shyness. *Journal of Personality*, **45**, 596–611.

Pirsig, R. M. (1974). *Zen and the art of motor cycle maintenance: an enquiry into values*. (London: Bodley Head).

Plum, T. (1977). Fooling the user of a programming language. *Software Practice and Experience*, **7**, 215–222.

Poulson, D. F., (1983). *An investigation of cognitive factors relating to the way that people respond to the use of computers*. Unpublished PhD thesis, Loughborough University.

Powell, B. (1981). *Overcoming shyness*. (New York: McGraw-Hill).

Rauta, I. and Hunt, A. (1972). *Fifth form girls: their hopes for the future*. (London: Office of Population Censuses and Survey, HMSO).

Report on the future of broadcasting (Lord Annan, chairman). (1977). Cmnd. 6753; 6753–1 vi. (London: HMSO).

Reynolds, C. (1982). Code junkies hooked on micros. *Sinclair User*, December, 42–43.

Roe, A. (1953). A psychological study of eminent psychologists and anthropologists, and a comparison with biological and physical scientists. *Psychological Monographs*, **67**, 352, 1–54.

Roe, A. (1957). Early determinants of vocational choice. *Journal of Counselling Psychology*, **4**, 212–217.

Roe, A. (1957). Early differentiation of interests. In *Second research conference on the investigation of creative scientific talent*, edited by W. Taylor and A. Roe,

(Utah: University of Utah), pp. 98–108.
Roe, A. and Siegelman, M. (1964). A study of the origin of interests. *American Personnel Guidance Association Inquiry Study*, **1**, 1–98.
Rosner, S. and Abt, L. E. (1970). *The creative experience*. (New York: Dell Publishing Inc).
Rothery, B. (1971). *The myth of the computer*. (London: Business Books Ltd).
Runkel, P. J. (1956). Cognitive similarity in facilitating communication. *Sociometry*, **19**, 178–191.
Russell, C. A. (1976). *Science and the rise of technology since 1800*. Unit 4, O.U. AST 281. (Milton Keynes: Open University Press).
Russell, J. (1986). *The influence of gender on the performance of 7 year olds carrying out computer based tasks*. Unpublished BSc project report, Loughborough University.
Rutherford, E. E. and Mussen, P. H. (1968). Generosity in nursery school boys. *Child Development*, **39**, 755–765.
Sandall, P. D. (1960). *An analysis of the interests of seconday school pupils*. Unpublished PhD thesis, University of London.
Sandall, P. D. (1967). *Manual of instructions of the factoral interest blank*. National Foundation of Educational Research. London.
Santrock, J. W. (1970). Influence of onset and type of paternal absence on the first four Eriksonian development crises. *Developmental Psychology*, **3**, 273–274.
Saunders, F. E. (1975). Sex roles and the school. *Prospects, Quarterly Review of Education*, **5**, 362–371.
Saville, P. (1972). *The British standardisation of the 16PF supplement of norms. Forms A and B*. (London: NFER-Nelson).
Saville, P. (1978). *A critical analysis of Cattell's model of personality*. Unpublished PhD thesis, Brunel University. Microfilm No. D7749/78, British Lending Library.
Schachter, S. (1963). Birth order, eminence and higher education. *American Sociological Review*, **28**, 757–768.
Schiebe, K. E. and Erwin, M. (1979). The computer as alter. *Journal of Social Psychology*. 108, 103–109.
Schwartz, M. D. (1983). The impact of computer mediated work on individuals and organizations. *Computing, Psychiatry and Psychology U.S.A.* **5**, 6–8.
Shallis, M. (1984). *The silicon idol: the micro revolution and its social implications*. (Oxford: Oxford University Press).
Shore, J. E. (1985). *The sachertorte algorithm and other antidotes to computer anxiety*. (New York: Viking Penguin).
Siegelman, M. (1970). Loving and punishing parental behaviour and introversion tendencies in sons. In *Readings in extraversion-introversion*, edited by H. J. Eysenck, (London: Staples Press).
Simons, G. (1985). *Silicon shock: the menace of the computer invasion*. (Oxford: Basil Blackwell).
Simons, G. (1985). *The biology of computer life: survival, emotion and free will*. (Brighton: The Harvester Press).
Simpson, B. (1983). Why kids love chips. *The Guardian*, 15th December.
Skinner, B. F. (1954). The science of learning and the art of teaching. *Harvard Educational Review*, **24**, 2, 86–97.

Smail, B. (1984). *Girl friendly science: avoiding sex bias in the curriculum*. Schools Council Publications (York: Longman Resources Unit).
Smail, D. (1984). *Illusion and reality*. (London: J. M. Dent and Sons).
Solano, C. H., Batten, P. G. and Parish, E. A. (1982). Loneliness and patterns of self-disclosure. *Journal of Personality and Social Psychology*, **43**, 524–521.
Spelke, E., Zelazo, P., Kagan, J. and Kotelchach, M. (1973). Father interaction and separation protest. *Development Psychology*, **9**, 83–90.
Sprandel, G. (1982). A call to action—psychological impacts of computer usage. *Computers and Society (USA)*, **12**, 2, 12–13.
SPSS. Inc. (1983). *User's guide: a complete guide to SPSSx language and operations*. (New York: McGraw-Hill).
Stanworth, M. (1981). *Gender and schooling: a study of sexual divisions in the classroom*. (London: Women's Research and Resources Centre).
Stewart, T. F. M. (1981). User friendly systems. Presented to *Butler Cox Foundation 9th Management Conference*, Bournemouth.
Stewart, T. F. M. (1984). More practical experiences in solving VDU ergonomic problems. In *Health hazards of VDTs?* edited by B.G. Pearce (Chichester: John Wiley).
Strong, E. K. (1943). *Vocational interests of men and women*. (Stanford, CA: Stanford University Press).
Suedfeld. P. (1964). Birth order of volunteers for sensory deprivation. *Journal of Abnormal and Social Psychology*, **68**, 195–196.
Surtees, J. (1988). Memories are made of this. *The Sunday Times*, 31st July, C11.
Sylva, K., Bruner, J. S. and Genova, P. (1976). The role of play in problem-solving of children 3-5 years old. In *Play: its role in development and evolution*, edited by J. S. Bruner, A. Jolly and K. Sylva, (Harmondsworth: Penguin).
Thimbleby, H. (1978). Dialogue determination. Paper presented at the INSPEC Seventh Cranfield Conference on Mechanised Information and Retrieval Systems.
Thimbleby, H. (1979). Computers and human consciousness. *Computers and Education*, **3**, 241–243.
Thomas, F. T. and Young, P. T. (1938). Liking and disliking persons. *The Journal of Social Psychology*, **9**, 169–187.
Thorndike, E. L. (1935). The interests of adults. *Journal of Educational Psychology*, **26**, 497–507.
Tinsley, H. E. A., Barrett, T. C. and Kass, R. A. (1977). Leisure activities need satisfaction. *Journal of Leisure Research*, **9**, 110–120.
Tittscher, R. (1984). *The UK microcomputer market: major companies and trends*. (London: Sector Investments Ltd).
Tjonn, H. H. (1984). Report of facial rashes among VDU operators in Norway. In *Health hazards of VDTs?*, edited by B. G. Pearce, (Chichester: John Wiley).
Toffler, A. (1971). *Future Shock*. (London: Pan Books).
Tseng-Wen-Sheng, (1973). Psychopathological study of obsessive-compulsive neurosis in Taiwan. *Comprehensive Psychiatry*, **14**, 139–150.
TUC Guidelines on VDUs. (1986). (London: Trades Union Congress).
Tuckman, B. W. (1984). Thinking out loud—Why (and why not) teach computer usage. *Educational Technology*, February, 35.
Turkle, S. (1984). *The second self: computers and the human spirit*. (London: Granada).

Valentine, C. W. (1956). *The normal child and some of his abnormalities.* (Harmondsworth: Pelican).
Vallee, J. (1984). *The network revolution: confessions of a computer scientist.* (Harmondsworth: Penguin).
Vernon, P. E. (1949). Classifying high-grade occupational interests. *Journal of Abnormal and Social Psychology*, **44**, 85–96.
Vidler, D. C. (1977). Achievement motivation. In *Motivation in Education*, edited by S. Ball, (New York: Academic Press).
Waddilove, K. (1984). The case for cost effective chalk. *The Guardian*, 20th March, 11.
Warburton, F. W. (1968). *The relationship between personality and scholastic attainment.* University of Manchester. Cited in Entwistle (1972), *British Journal of Education Psychology*, **42**, 137–151.
Ward, R. (1983). Wasted potential. *Times Educational Supplement*, November.
Ward, R. (1985). Girls and computing. *Computer Education*. February, 4–5.
Warren, J. R. (1963). The effects of certain selection procedures. *Special Report No. 8*. (Nebraska: University of Nebraska).
Weeks, D. J. (1988). *Eccentrics: the scientific investigation*. (Stirling: Stirling University Press).
Weinberg, G. M. (1971). *The psychology of computer programming*. (New York: Van Nostrand Reinhold).
Weinberg, S. (1982), cited by R. Wrege in High (Tech) Anxiety. *Popular Computing*. January, 46–52.
Weiner, G. (1980). Sex differences in mathmatical performance. In *Schooling for Women's Work*, edited by R. Deem, (London: Routledge and Kegan Paul).
Weiss, F. A. (1962). Self alienation: dynamics and therapy. In *Man alone: alienation in modern society*, edited by E. Josephson and M. Josephson, (Burnley: Dell Publishing).
Weizenbaum, J. (1976). *Computer power and human reason*. (San Francisco, CA: W. H. Freeman and Company).
Weizenbaum, J. (1984). *Computer power and human reason*. (Harmondsworth: Penguin).
Winterbottom, M. (1958). The relation of need for achievement to learning experiences in independence and mastery. In *Motives in fantasy, action and society*, edited by J. W. Atkinson, (Princeton, NJ: Van Nostrand).
Witkin, H. A. (1950). Individual differences in ease of perception of embedded figures. *Journal of Personality*, **19**, 1–15.
Witkin, H. A., Dyk, R. B., Faterson, H. F., Goodenough, D. R. and Karp, S. A. (1962). *Psychological differentiation*. (New York: John Wiley).
Work Research Unit. (undated). *On Improving the quality of working life*. (London: Department of Employment).
World Health Organization. (1964). *Evaluation of dependence-producing drugs*. World Health Organization Technical Report Series No. 287, Geneva.
World Health Organization. (1964). *Expert committee on drug dependence*. World Health Organization Technical Report Series No. 116, Geneva.
Wright, S. (1955). *Some psychological and physiological correlates of certain academic underachievers*. Unpublished PhD thesis, University of Chicago. Cited by Cattell *et al.* (1970), *Handbook for the 16PF Questionnaire*, (Champaign, IL: IPAT).

Wylie, H. L. amd Delgado, R. A. (1959). Pattern of mother-son relationship involving the absence of the father. *American Journal of Orthopsychiatry*, **29**, 644–649.

Zarat, M. M. (1984). Cataracts and visual display terminals. In *Health hazards of VDTs?*, edited by B. G. Pearce, (Chichester: John Wiley).

Zimbardo, P. G. (1977). *Shyness*. (Reading, MA: Addison-Wesley).

Zimbardo, P. G. (1980). The age of indifference. *Psychology Today*, August, 71–76.

Zimbardo, P. G. (1980). The hacker papers. *Psychology Today*, August, 62–70.

Zimbardo, P. G. (1981), cited by D. Ingber. Computer Addicts. *Science Digest*, July, 88–91 and 114–115.

Zimbardo, P. G. (1981). *Shyness: What it is and what to do about it*. (London: Pan).

Author Index

Subject Index

Page numbers in italics refer to Tables and Figures